Research Paradigm Considerations for Emerging Scholars

Full details of all our other publications can be found on http://www.channelviewpublications.com, or by writing to Channel View Publications, St Nicholas House, 31–34 High Street, Bristol BS1 2AW, UK.

Research Paradigm Considerations for Emerging Scholars

Edited by
Anja Pabel, Josephine Pryce and Allison Anderson

CHANNEL VIEW PUBLICATIONS
Bristol • Blue Ridge Summit

DOI https://doi.org/10.21832/PABEL8274
Library of Congress Cataloging in Publication Data
A catalog record for this book is available from the Library of Congress.
Names: Pabel, Anja, editor. | Pryce, Josephine, editor. | Anderson, Allison, editor.
Title: Research Paradigm Considerations for Emerging Scholars/Edited by
   Anja Pabel, Josephine Pryce and Allison Anderson.
Description: Bristol, UK; Blue Ridge Summit, PA : Channel View Publications,
   2021. | Includes bibliographical references and index. |
Summary: "This book provides insights into the experiences and reflections of
   researchers as they negotiate the world of paradigms and seek to find their niche.
   It offers an accessible exploration of research paradigms and will be a valuable
   resource for postgraduate researchers, emerging scholars and PhD supervisors"—
   Provided by publisher.
Identifiers: LCCN 2020056169 (print) | LCCN 2020056170 (ebook) |
   ISBN 9781845418267 (paperback) | ISBN 9781845418274 (hardback) |
   ISBN 9781845418281 (pdf) | ISBN 9781845418298 (epub) | ISBN 9781845418304
   (kindle edition)
Subjects: LCSH: Tourism—Research—Methodology. | Research—Methodology.
Classification: LCC G155.A1 R4739 2021 (print) | LCC G155.A1 (ebook) |
   DDC 306.4/819—dc23 LC record available at https://lccn.loc.gov/2020056169
LC ebook record available at https://lccn.loc.gov/2020056170

British Library Cataloguing in Publication Data
A catalogue entry for this book is available from the British Library.

ISBN-13: 978-1-84541-827-4 (hbk)
ISBN-13: 978-1-84541-826-7 (pbk)

Channel View Publications
UK: St Nicholas House, 31–34 High Street, Bristol, BS1 2AW, UK.
USA: NBN, Blue Ridge Summit, PA, USA.

Website: www.channelviewpublications.com
Twitter: Channel_View
Facebook: https://www.facebook.com/channelviewpublications
Blog: www.channelviewpublications.wordpress.com

Copyright © 2021 Anja Pabel, Josephine Pryce, Allison Anderson and the authors of individual chapters.

All rights reserved. No part of this work may be reproduced in any form or by any means without permission in writing from the publisher.

The policy of Multilingual Matters/Channel View Publications is to use papers that are natural, renewable and recyclable products, made from wood grown in sustainable forests. In the manufacturing process of our books, and to further support our policy, preference is given to printers that have FSC and PEFC Chain of Custody certification. The FSC and/or PEFC logos will appear on those books where full certification has been granted to the printer concerned.

Typeset by Nova Techset Private Limited, Bengaluru and Chennai, India.
Printed and bound in the UK by the CPI Books Group Ltd.
Printed and bound in the US by NBN.

Contents

	Figures and Tables	vii
	Contributors	ix
1	Embarking on the Paradigm Journey *Anja Pabel, Josephine Pryce and Allison Anderson*	1
2	Logical Positivism in Consumer Behaviour Research *Antje R.H. Graul*	12
3	The Design Science Research Paradigm: An Instantiation of Website Benchmarking *Leonie Cassidy*	25
4	An Application of Quasi-Experiments to Study Humour in Tourism Settings Guided by Post-Positivism *Anja Pabel*	38
5	Knowledge Co-Production in Tourism and the Process of Knowledge Development: Participatory Action Research *Tramy Ngo, Gui Lohmann and Rob Hales*	50
6	The Constructivist Paradigm and Phenomenological Qualitative Research Design *Justyna Pilarska*	64
7	Applying the Interpretive Social Science Paradigm to Research on Tourism Education and Training *Yohei Okamoto*	84
8	An Autoethnographic Chronicle on the Ethnographic Exploration of the Nature of Hotel Work and Hospitality in Far North Queensland *Josephine Pryce*	97
9	Neo-Tribalism through an Ethnographic Lens: A Critical Theory Approach *Oscar Vorobjovas-Pinta*	112
10	Navigating the Complex Variety of Feminisms *Linda Colley and Sue Williamson*	130

11 The Pragmatic Paradigm in Destination Competitiveness
    Studies: The Case of the SCUBA Diving Tourism Niche   142
    *Ambrozio Queiroz Neto, Gui Lohmann, Noel Scott
        and Kay Dimmock*

12 Pragmatism in the Context of Urban Design and Tourism:
    A Multidisciplinary Study   158
    Allison Anderson

13 In Search of an Intermediate Paradigmatic Ground: Critical
    Realism-Post-Positivism in Understanding Tourists'
    Motivation and Experiences in Asian Spas   171
    Jenny H. Panchal

14 Paradigms: A Supervisor's Perspective and Advice   189
    Philip L. Pearce

15 Into the Future: Moving Forward with Reflective Practice on
    Paradigms   203
    Josephine Pryce

    Index   213

Figures and Tables

Figures

Figure 1.1	The research design framework showing the interconnection of paradigm, research methodology and research methods	7
Figure 3.1	Design science research (DSR) staging of artefact solution	29
Figure 3.2	Design science research methodology (DSRM) for website analysis method (WAM)	31
Figure 3.3	The website analysis method (WAM) artefact	32
Figure 3.4	Design science research methodology (DSRM) steps to constructing my thesis	34
Figure 5.1	Knowledge development process included in the research design	57
Figure 7.1	A philosophically minded approach to tourism research	85
Figure 9.1	Neo-tribal characteristics	120
Figure 11.1	Identification on knowledge gap	147
Figure 11.2	Research framework	149
Figure 11.3	Methodological framework	153
Figure 12.1	The changing focus of disciplinary research	161
Figure 13.1	The differences between qualitative and quantitative research paradigms	183

Tables

Table 2.1	Factorial design	20
Table 4.1	Data collection at the two tourism attractions	45
Table 5.1	Stages of the PAR process	54
Table 5.2	Background information of the investigated CBTEs	56
Table 6.1	The researcher's steps and activities in the modified Stevick–Colaizzi–Keen method as described by Moustakas (1994)	76
Table 6.2	Composite descriptions representing the essence of the experience of the interviewees	78
Table 9.1	Applications of neo-tribalism in tourism research	119
Table 11.1	Analysis of important destination competitiveness attributes	146
Table 11.2	Profile of respondents in stage one	151
Table 12.1	Pragmatic alternative to the key issues in social science research methodology	162
Table 12.2	Research approach	164
Table 13.1	The major research paradigms	178
Table 13.2	The current research paradigm	181

Contributors

Allison Anderson has spent her working life spanning the academic and professional spheres of tourism and strategic planning, which explains much of her pragmatic research philosophy. She started her academic career in earth sciences, gaining a Bachelor of Science (honours) at Victoria University of Wellington, NZ. After building a consulting career advising on tourism development in Tasmania and North Queensland, she completed a Master of Urban & Regional Planning at James Cook University in Cairns, receiving an academic medal for her efforts. She worked as a strategic planner in an engineering firm, before completing her PhD in urban design and tourism, with the help of a Queensland Smart State PhD scholarship. Just before she submitted her PhD, she was appointed as a Lecturer in Tourism Planning at CQUniversity in Cairns, which then led to her current role leading the research and insights programme at Tourism Tasmania. Allison is also an Associate at the University of Tasmania.

Leonie Cassidy has a multidisciplinary PhD from James Cook University, Australia. Her research interests lie in the areas of climate change, coral bleaching, the Great Barrier Reef, tourist behaviour (including culture, digital device use, website use, photography and social media usage), website design (benchmarking, best practice and quality) and design science research. Leonie has published several journal articles, book chapters and articles in refereed proceedings in these areas, and continues research activities with CQUniversity, Cairns campus. She also lectures part time at James Cook University, including developing subjects in the areas of technopreneurship and entrepreneurship.

Linda Colley is an Associate Professor of Human Resources/Industrial Relations at CQUniversity, based on the Brisbane campus and affiliated with the Appleton Institute. She brings extensive practical experience from her career in HRM and industrial relations in the Queensland public service. Her research builds on this career, including both historical study of public employment trajectories and contemporary challenges such as workforce planning and institutional reform. She has published on topics such as gender, merit, tenure, job security, gender and age at work, public management reform and the effects of austerity measures on public employment.

Kay Dimmock, PhD is a Senior Lecturer at Southern Cross University based in the School of Business and Tourism. Her work extends from business to tourism in higher education. In a tourism context Kay has researched social experiences including perceptions, beliefs and impacts on stakeholders through leisure and marine tourism. Her research has been individual and collaborative at local and regional levels. She works with tourism stakeholders including industry, community, researchers and government agencies to understand social dimensions of coastal and marine environments. Internationally, Kay works with fellow collaborators and has co-edited a textbook focused specifically on the tourism niche, *SCUBA Diving Tourism*. Supervision and mentoring HDR students to successful completion in marine related topics has involved Kay with projects including 'Phenomenology, women and surfing', 'Destination competitiveness and SCUBA diving tourism', 'White shark cave diving in Australia and potential for conservation' and 'Wreck divers and management of underwater cultural heritage'.

Antje R.H. Graul, PhD, is an Assistant Professor of Marketing at the Jon M. Huntsman School of Business at Utah State University. Her research interests lie in the area of consumer decision-making and behaviour regarding sustainable modes of consumption such as consumer sharing and electric vehicle adoption. Before following her research interest, she spent two years in the industry focusing on marketing communications and public relations at the BMW Group, Germany.

Rob Hales is the Director of the Griffith Centre for Sustainable Enterprise in the Griffith Business School. Rob is also the Program Director of the Master of Global Development and teaches courses in leadership for sustainability and research methods. His research interests include climate change policy, sustainable business, sustainable tourism and indigenous studies. He is also involved in governance of the sustainable operations of Griffith University, Queensland. His background in running a small business in tourism and environmental education informs his research and teaching.

Gui Lohmann is Professor in Air Transport and Tourism Management at Griffith University (Brisbane, Australia), one of the highest-ranked universities globally for tourism studies. He has authored several books, including editing *Tourism in Brazil: Environment, Management and Segments* and co-authoring *Tourism Theory: Concepts, Models and Systems*, published in English, Spanish and Portuguese. He has published broadly internationally, including more than 40 journal articles in leading peer-reviewed tourism publications, such as *Annals of Tourism Research, Journal of Sustainable Tourism, Current Issues in Tourism, Journal of Destination Marketing and Management* and *International Journal of*

Hospitality Management. He is also on the Editorial Board of *Tourism Review* and the *Journal of Air Transport Management*. At Griffith University, Gui has successfully supervised 10 PhD scholars from countries as diverse as Brazil, China, New Zealand, Serbia and Vietnam.

Tramy Ngo holds a PhD in Tourism Management from Griffith University (Australia). She is a tourism expert with a strong interest in tourism impact assessment, community development in tourism, Indigenous tourism, sustainable tourism and cross-cultural studies. After over 10 years working as a tourism lecturer in Vietnam, Tramy currently spans her professional expertise on the areas of tourism research and consultancies in Australia. Tramy has published her research projects in leading journals such as *Journal of Sustainable Tourism* and *Current Issues in Tourism*. The most recent consultation projects she contributed to include the Queensland Wine Tourism Strategy, the Central Highlands (Queensland) Caravan & Camping Strategy, and the Moreton Bay Region Tourism Economic Impact Analysis.

Yohei Okamoto (MPhil) is an Associate Lecturer in Tourism and Events at Murdoch University in Perth Australia. Yohei's current research focuses on philosophical issues in post-secondary tourism education, and digital game-based learning (DGBL) as a pedagogical tool for sustainable tourism education. He has also worked on research projects relevant to trails, community festivals and bicycle tourism. He is about to embark on his PhD journey on global nomads on bicycles.

Anja Pabel is a lecturer in tourism at CQUniversity in Cairns, Australia. Her research interests include tourist behaviour, marine tourism, tourism sustainability and humour research. Anja completed a Bachelor of Business with first class honours in 2009 at James Cook University (JCU). In 2015, she completed her PhD in tourism at JCU. Her PhD research investigated the role of humour in tourism settings. More specifically she examined in what ways humour influences the tourist experience in making tourists feel comfortable, connected and more mindful.

Jenny H. Panchal is a Senior Lecturer in the College of Law, Business and Governance at James Cook University (Townsville, Australia). She received the Australian Leadership Award (ALA) Scholarship which funded her PhD in Tourism at JCU. She was also an NZAID Scholarship Awardee which enabled her to obtain her Masters in Tourism Management from Victoria University of Wellington, New Zealand. At JCU, she's been involved in teaching tourism, events management and other management subjects in the undergraduate and postgraduate levels in Singapore and Australia, both internally and externally. Dr Panchal's research interests broadly include tourist behaviour related to specific forms of tourism (e.g.

wellness and spa tourism, boutique hotels, perceptions of poverty), positive psychology, and are notably focused on the Asian context. Her involvement in Southeast Asian research includes linkages to the ASEAN Tourism Research Association (ATRA) as the Deputy Secretary.

Philip L. Pearce was the Foundation Professor of Tourism and a Distinguished Professor at James Cook University, Australia. He wrote and edited 18 tourism books and has around 300 publications. His longstanding interests were studies in tourist behaviour and experience. He was well known for his work on travel motivation, and approaches to tourists' experience as well as special topics including humour stemming from positive psychology. He worked with his PhD students and international colleagues using a variety of methods and approaches. Most studies were done in settings in Asia, Australia and Europe.

Justyna Pilarska holds a PhD in Social Sciences. She joined the Faculty of Historical and Pedagogical Sciences at the University of Wroclaw in 2012 as an Assistant Professor at the Department of Intercultural Education and Social Support Research, Institute of Pedagogy, where she teaches methodology of social research, contemporary European culture, the Balkans in the 20th and 21st centuries, as well as a number of other subjects. Although she works broadly in intercultural pedagogy, she has published articles in anthropology, sociology, urban pedagogy and cross-cultural education, thus much of her work crosses disciplinary boundaries. Her research interests lie at the intersection of Bosnian cultural diversity, religious syncretism in Japan, multiculturalism and methodology of social research, including indigenous methodologies.

Josephine Pryce is an Associate Professor at James Cook University, Australia. She is a scholar of Organisational Behaviour with a keen interest in 'the nature of work' and research methodologies. She came to academia with a strong industry background having worked for Australia Post and the hotel industry, the latter in North Queensland. During that time, she developed a strong appreciation of the challenges faced by workers, employers and organisations in service industries. Her PhD extended understanding of service predispositions, human relations in the workplace, and organisational culture. Josephine seeks to enrich her perspectives through research that leads to better understanding of the lives of people in communities and the world, especially in relation to their working lives. She is also keenly interested in paradigms, ontologies, epistemologies and how these inform methodological approaches.

Ambrozio Correa de Queiroz Neto, PhD, is Associate Professor in Tourism Management at the Department of Tourism, Centro Federal de Educação Tecnológica Celso Suckow da Fonseca – CEFET-RJ in Brazil.

His research interests focus on tourism destination management, destination competitiveness, tourism planning and SCUBA diving tourism.

Noel Scott, PhD, is Adjunct Professor of Tourism Management, in the Sustainable Research Centre, University of Sunshine Coast, Queensland, Australia. His research interests include the study of tourism experiences, and destination management and marketing. He is a frequent speaker at international academic and industry conferences. He has over 300 academic articles published including 16 books. He has supervised 23 doctoral students to successful completion of their theses. He is on the Editorial Board of 10 journals, a member of the International Association of China Tourism Scholars and a Fellow of the Council for Australasian Tourism and Hospitality Education. Prior to starting his academic career in 2001, Dr Scott worked as a senior manager in a variety of businesses including as Manager of Research and Strategic Services at Tourism and Events Queensland.

Oscar Vorobjovas-Pinta is a leading expert on LGBTQI+ communities in the context of leisure, hospitality and tourism. His research interests are the sociology of tourism, tourist behaviour and LGBTIQ+ tourism. Dr Vorobjovas-Pinta explores LGBTIQ+ travellers as neo-tribes, who come together from disparate walks of life but are united through shared sentiment, rituals and symbols. He has a broad interest and industry experience in innovative tourism research, and has been involved in a number of projects focusing on resilience, and on technology enabled advanced tracking of visitors to Tasmania and Sydney, New South Wales. These projects have each been built on extensive stakeholder and industry engagement.

Sue Williamson is a Senior Lecturer in Human Resource Management at the University of New South Wales, Canberra. She specialises in two main areas of research – gender equality in the workplace, and public sector human resources and industrial relations. She is currently examining how public sector organisations can create and sustain gender equality, and inclusive cultures. Sue was chief investigator, leading a consortium of researchers to examine how middle managers are progressing gender equality, funded by the Australia and New Zealand School of Government. Sue has published widely on this topic and also shares her findings with industry partners and the community.

1 Embarking on the Paradigm Journey

Anja Pabel, Josephine Pryce and Allison Anderson

Introduction

This book aims to be a timely and valuable resource for you, as a higher-degree research (HDR) scholar, as you explore the paradigm landscape and seek to discover and engage with the paradigm that aligns with you as a researcher on your PhD journey. It provides a platform to learn about the paradigm journeys of other emerging scholars and to demonstrate 'paradigms in action'.

As HDR scholars engaging with our research, we are often faced with trying to understand the complexities and ambiguities of the literature pertaining to various paradigms, ontologies and epistemologies. As emerging scholars, we may find it difficult to understand the 'heavy conceptual constructs and philosophical underpinnings of research paradigms' (Killion & Fisher, 2018: 9). The literature on paradigms can be contentious, and contain contradictory and emerging ideologies, making the journey towards understanding the philosophical underpinnings of these domains difficult and confusing. Hence, determining the paradigm that best aligns with our personal ideologies as scholars and the underlying assumptions of the research can be rather daunting, challenging and perplexing, especially as novice researchers. Moreover, we may not always find ourselves in environments where collegial support and insight are available.

This book provides you as HDR candidates and emerging scholars with information about the experiences and reflections of other scholars as they sought to understand their paradigm and its influence on their approach to research. Hence, the book is oriented mainly towards postgraduate HDR candidates and emerging scholars who are curious about paradigms and have questions about paradigmatic issues. A secondary target audience is higher-degree supervisors or 'lecturers of research courses' seeking to enhance their understanding of a specific paradigm in their supervisory activities or who might like to use this text as a resource for their students. For all readers, the book presents insights into the lived experiences of researchers as they have journeyed through the undulating terrain of exploring paradigms.

A Brief 'Historical' Context of the ROPE Group

In retrospect, PhDs generally look masterfully crafted to perfectly capture the most meaningful insights and contributions to theory and practice in the world of knowledge. But the entangled and challenging pathways to reach that end point are never evident at the beginning. One of the early milestones in the PhD journey is the 'confirmation-of-candidature'. This milestone involves preparation of a proposal outlining the research design and presentation to a panel of experts who evaluate the candidate's capability in knowledge and skills. To achieve confirmation-of-candidature, we are required to articulate our research objectives and demonstrate our plans for building a research project to achieve those objectives. It is often only after we have been successful in our confirmation-of-candidature, officially installed as HDR scholars in the university, and attended induction courses, that supervisors will sit down and suggest that we consider our philosophical framing. It is then that questions of disbelief arise, and we begin to ask: What is philosophical framing? And do I actually need one, or can I just get on with the project?

Some authors in this tome would say, 'you can take the easy road or the hard road on this dilemma', but most candidates will eventually need to determine a philosophical framing for their research. Allison Anderson (co-editor of this book) realised early on that what she was missing was a substantial education in disciplinary structures and their philosophical underpinnings. In response to her lack of understanding about all things paradigmatic, Allison started talking with friends who were also doing their PhDs, such as co-editor Anja Pabel. In truth, there were many of us who were rather vague about the process of working out an understanding of the paradigmatic underpinnings of our research. Gathering together a core group of around 10 PhD candidates from a diverse range of disciplines across the university, we formed a group we called 'ROPE' – the group was to talk about Research, Ontologies, Paradigms and Epistemologies.

Collectively the ROPE group agreed that we were the 'blind leading the blind' without a guide. We needed someone who had knowledge and understanding of all matters relating to paradigms. Anja approached Jo Pryce (co-editor) to help guide us through our fortnightly discussions. It transpired that Jo was very interested in paradigms and enjoyed conversing on 'all things paradigmatic'. This strategy proved to be critical to our success as a group with an ongoing interest in advancing conversations on paradigms and their underpinnings. Jo provided us with the ultimate structure through which we could explore the paradigms and lead discussions about our research. A watershed moment was when Jo introduced us to a table from notable authors Guba and Lincoln (1994, 2005). The table highlights the main paradigms and illustrates the association between paradigms, ontologies and epistemologies. As we worked through the diagram, each member of ROPE was able to recognise the paradigm and ontology that related to their research. Consequently, it was decided that at each meeting one of us would deliver a presentation on our respective research and lead a discussion for that meeting on the associated paradigm's advantages and disadvantages. We regularly invited experienced academics who were experts in specific paradigms or methodologies to guide us through the paradigmatic maze and to help us in making sense of our thoughts and discussions.

The ROPE meetings where open to any HDR scholars on campus, and our discussions were open to anything pertaining to research but predominantly aimed to create a supportive culture around the research process. The group was a safe space where we felt comfortable to ask 'dumb questions' and to share the triumphs and struggles of our respective research journeys. We also had open discussions around the realities and challenges HDR scholars often face in dealing with supervisors with limited time and provided emotional support to each other in our networks to better cope. ROPE lasted for about three years, essentially until that cohort of higher-degree research candidates completed their PhDs and moved on to secure employment.

It was from the ROPE group that the idea of this book was born. We were aware that every HDR scholar had a unique story to share with their own struggles. We thought that a book that outlines these journeys would be of value to other emerging scholars. When we commenced working on a book proposal, invitations were sent to potential authors. Most of the contributing authors were still in the midst of writing their PhD theses and living through the real issues faced by HDR scholars when deciding on what paradigm and research process were most suitable to answer their proposed research questions. Hence this book offers fresh perspectives of these journeys and makes this book a valuable educational resource.

Structure of Each Chapter

In this first chapter, we provide an introduction and brief descriptions on key paradigmatic terminology. Hence, while readers may target specific chapters or paradigms, it is advisable to read this first chapter along with the chapter of interest. Each chapter is presented by an author who explains the specific philosophical framework in the context of their research topics; for example tourism, consumer behaviour, organisational behaviour, website benchmarking and women at work.

Each chapter includes the following sections:

- *A review of the paradigm and general description of its tenets.* This will include historical and philosophical information as well as dominant theories and methods associated with the paradigm. This section will be useful for readers as a resource/reference source on where to find more information on a specific paradigm.

- *Application of the paradigm.* This part will describe how the author(s) applied the paradigm in the specific context of their own research. This will include a brief overview of the author's research topic and describe and discuss the nature of their collected and analysed data, explaining how the paradigm is linked to its epistemology and methodology. This will be useful for readers to understand how paradigms apply to a particular research context.
- *Reflection on the author's use of the paradigm.* This part focuses on reflexivity, where the author is reflecting on their choice and use of a specific paradigm and includes personal anecdotes and experiences in working within it. This section will be particularly useful for HDR scholars to recognise some of the advantages or limitations of each paradigm. It is this final and reflexive section of each chapter that offers insights into the journeys taken by the authors of this book.

We envision that after reading a particular chapter or a series of chapters from this book, you will gain relevant knowledge about a single or several paradigm(s) and have a deeper understanding of how it or they can be applied to a specific research project. In addition, the reflections provided by the authors of each chapter lend insight into the ways in which they as emerging scholars came to understand and work with their paradigms. We hope that the reflexive nature of this book will provide you with guidance for the projects you are designing and/or proposing to undertake. According to Robson (2002: 22), reflexivity is about becoming aware 'of the ways in which the researcher as an individual with a particular social identity and background has an impact on the research process'. Being reflexive about our research allows us, as investigators, to acknowledge, recognise and accept understandings of the issues under investigation which is helpful in the creation of knowledge (Killion & Fisher, 2018). Reflexivity, it has been argued, allows for an enhanced understanding of personal, intersubjective, and social processes that may influence the investigation and rests on the awareness of self (Cutcliffe, 2003). It is this reflexive approach that differentiates this book from others in the paradigm area and provides a unique contribution to the experience of researchers, especially emerging scholars.

Defining What a Paradigm is

No research takes place in a vacuum; rather all research studies are shaped and moulded within the parameters of the chosen research paradigm (Killion & Fisher, 2018). Let us start by exploring some of the multiple definitions which have been developed to explain what a paradigm is and what it entails. According to Kuhn (1962) a paradigm represents a particular way of thinking that is shared by a community of scientists in solving problems in their discipline. As such, our worldviews, as HDR scholars, are likely to be moulded by the discipline area in which we operate, and the paradigmatic beliefs of our supervisory panel (Creswell, 2014).

Paradigms are also often described as a 'school of thought' (Sousa *et al.*, 2007: 503) which hold 'a basic set of beliefs that guides action' (Guba, 1990: 17). Paradigms provide a general orientation about the world and are also linked to certain 'accepted practices of a research community in a particular historical time and context' (Munar & Jamal, 2016: 2). For McGregor and Murnane (2010), paradigms are indicative of the intellectual integrity, trustworthiness and diversity of scholarship which depend on researchers accounting for the philosophical underpinnings, concepts and values of their work. Mertens (2005: 194) explains that a paradigm 'influences the way knowledge is studied and interpreted.' Fundamentally, paradigms serve as frameworks that help us shape 'what should be studied, what is seen and how what is seen is interpreted or understood' (Killion & Fisher, 2018: 11).

Nowadays, as HDR scholars, we are expected to outline the philosophical worldview assumptions in our theses. These paradigm sections include detailed sections on the philosophical worldview used for a research project and how the worldview shaped the chosen approach to research (Creswell, 2014). Reading about research paradigms helps us as HDR scholars to better understand the history behind knowledge formation in a given discipline; to be conscious of the common research processes used in a particular discipline; to apply the sense-making, identity-crafting and community-building tools; and to assess the level of maturity of a particular research field (Munar & Jamal, 2016). We are often relieved to find out that extensive paradigm sections are seldom required when presenting findings in journals or conference papers (Killion & Fisher, 2018). Some journals may still require a section on paradigmatic considerations, however, never to the extent that it is presented in a higher-degree research thesis (due to word limitations in most academic journals).

Numerous paradigms are outlined in this book, for example logical positivism, post-positivism, design science research (DSR) paradigm, pragmatism, constructivism, feminism, critical theory and multi-paradigmatic research. Each chapter is predominantly derived from the journey of recent PhD scholars or emerging scholars sharing their experiences with their chosen paradigm. Moreover, there are also more senior scholars reflecting on their use of paradigms and in one chapter, Professor Philip L. Pearce describes his experiences of guiding HDR scholars through the paradigmatic maze. It should also be noted that the paradigms mentioned in this edited book are based on social science topics (rather than a focus on natural sciences).

The Four Principles of Paradigms

Irrespective of whether we as HDR scholars find it a daunting and challenging experience to write about paradigms, we are nevertheless expected to be engaged in the dialogue as a proof that we understand the underlying assumption(s) of the chosen paradigm (Denzin & Lincoln,

2005). In this regard, it is helpful to understand that paradigms are frequently distinguished by four principles (Chilisa & Kawulich, 2012; McGregor & Murnane, 2010; Onwuegbuzie & Leech, 2005; Patton, 2002):

- *Ontology* – is concerned with questions around what constitutes reality, existence or being; that is, what do we believe about the nature of reality?
- *Epistemology* – considers questions around what counts as knowledge and truth, and how people come to know it, i.e. how do we know what we know and what sources of knowledge are reliable?
- *Axiology* – is concerned with questions around ethics and value systems; that is, what do we believe is true in terms of moral choices, ethics and normative judgements?
- *Methodology* – guides the researcher towards the use of appropriate approaches of enquiry; that is, how should we study the world?

For any emerging scholar it is useful to gain an understanding of what these four principles mean, because they help to make sense of the main assumptions of any paradigm. Grasping these four principles in the context of a particular paradigm is useful to 'frame a researcher's view of a research problem, how he/she goes about investigating it, and the methods he/she uses to answer the research questions' (Chilisa & Kawulich, 2012: 2). It should be acknowledged that none of the four principles function in isolation from the others. Instead, they are considered to be interrelated (Killion & Fisher, 2018). For example, underlying features of ontology impact directly on epistemology, methodology and axiology. When reading the chapters in this book, the main assumptions of the different paradigms should become clear, i.e. is there a single reality or are there multiple constructed realities, or is the research assumed to be value-free/objective or value-laden/subjective?

How to Choose the 'Correct' Paradigm?

Conversations about research do not usually start with a paradigm in mind. Research endeavours start with an idea or a research topic around which we as researchers develop a research aim or a research question. Once we have a research topic and aim in mind, there is the need to make decisions on how the topic will be investigated (Chilisa & Kawulich, 2012). These decisions may be informed by several factors including (Chilisa & Kawulich, 2012; Creswell, 2014; Killion & Fisher, 2018; Viglia & Dolnicar, 2020):

- Discipline-specific influence: What are the main assumptions, values and practices associated with viewing reality in your discipline? What constitutes credible findings in your discipline? What are the choices around methodology and research design that are principally used by a certain discipline? Who are the audience you are writing for?

- Supervisor influence: What are the paradigmatic suggestions made by your supervisor? Are you comfortable using the approaches suggested by your advisers?
- Personal considerations: How do you think about the research problem and how it can be studied? What is your own view on how knowledge is created and what constitutes reality? What is your own value system? These views will guide your thinking and assumptions about the world around you.

Keeping the entire research process in mind, it is important to recognise that no one paradigm is the 'correct one'. Your considerations should be mainly driven by what paradigm will help you to best answer the research question that you are trying to address in your study. The choice of a research paradigm can ultimately come to reflect the aims and purposes of the investigation (Killion & Fisher, 2018; Viglia & Dolnicar, 2020). By answering some of the above questions, you should be able to determine what paradigm aligns with your research question and select an appropriate methodology to start tackling your research project.

Research Design

The individual research design taken to achieve the aim of your project is another important consideration. According to Creswell (2014) the research design is the plan to conduct research which brings together considerations around the chosen paradigm and its assumptions about research, the specific research methodology and the research methods. This process highlights the interconnectivity between a paradigm and its associated ontology, epistemology and methodology (see Figure 1.1). Decisions around the research design should be driven by the research question(s) being investigated but it will also be influenced be your personal experiences with previous research designs you may have used. The three components of research design are outlined in Figure 1.1.

The paradigm section in a research thesis clearly describes any of the aspects outlined in Figure 1.1, starting with the overall paradigm and how

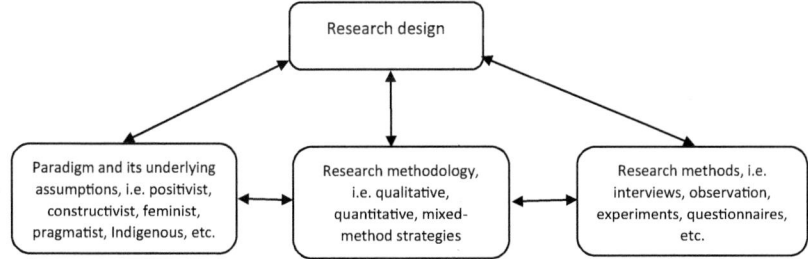

Figure 1.1 The research design framework showing the interconnection of paradigm, research methodology and research methods (Adapted from Creswell, 2014)

its philosophical assumptions influence the practice of research (Creswell, 2014). As mentioned before, your own worldview is likely to have an impact on how you are thinking about the topic and how you will approach a particular research project. It is important to be conscious of which paradigm and its related methodologies are being applied to your research topic because these considerations are going to influence how other scholars evaluate your research findings (McGregor & Murnane, 2010) and whether the conclusions drawn from your study can be perceived as valid and reliable (Viglia & Dolnicar, 2020). This process may even be iterative where you have settled on a paradigm but the underlying assumptions (ontologies and epistemologies) may be defined as you scope the research through the literature and give consideration to the practicalities of adopting particular methodologies.

It should also be noted that the terms 'methodology' and 'method' should not be used interchangeably since they relate to different aspects of the research process (McGregor & Murnane, 2010). Methodology deals with the overall operationalisation of an investigation; that is, relating to questions on how the overall investigation will be designed (Killion & Fisher, 2018). Methods, on the other hand, account for the more specific decisions and technical aspects of the research design, i.e. data collection techniques. The use of certain methods (e.g. questionnaires vs interviews) is frequently guided by the methodology (e.g. quantitative vs qualitative). Some research problems call for specific approaches (Creswell, 2014). For example, if the research problem calls for identification of factors and how they influence an outcome, or it calls for hypothesis testing, then a positivist/post-positivist paradigm might be most suitable. Alternatively, if the research problem requires a deeper understanding of the human mind and/or a social phenomenon, or if limited knowledge currently exists on a topic, then a constructivist/interpretivist paradigm is best. Each research design offers a unique approach or plan for answering a research question (Sousa *et al.*, 2007).

Multi-paradigmatic Research

There are interesting discussions in the literature on the so-called paradigm wars with their unending focus on two subcultures. On one end of the continuum is the subculture of positivist quantitative thinking, while on the other end lies the subculture driven by interpretivist qualitative thinking (Gage, 1989; Oakley, 1999). Treating the two orientations as bi-polar opposites has been criticised as 'divisive and, hence, counterproductive for advancing the social and behavioural sciences' (Onwuegbuzie & Leech, 2005: 268). Several scholars have argued that no research paradigm is superior to another because each has its own underlying assumptions and as such a specific way of producing unique knowledge (Alexander, 2006; Taylor & Medina, 2013).

Rather than focusing on ontological and methodological purism, as HDR scholar, you need to be able to appreciate the advantages and disadvantages of both orientations. This is because research tends to be more collaborative and long-term in nature these days, and it is the role of the research supervisors to teach their HDR candidates to be 'flexible in their investigative techniques, as they attempt to address a range of research objectives and questions that arise' (Onwuegbuzie & Leech, 2005: 290). Repeatedly it is being recognised that solving real problems in our dynamic world requires scholars at all levels to engage in collaborative interdisciplinary work and learn from the developments of such interactions and outcomes (Carpenter *et al.*, 2011). Hence there is a need for research courses to create awareness in HDR scholars about how to produce knowledge by using diverse research philosophies (Taylor & Medina, 2013); and, where possible, afford opportunities for HDR scholars to engage with research that is collaborative and inclusive of multi-paradigmatic approaches.

At large, this call has been answered by the notable increase in pragmatist approaches to addressing research questions, particularly in the social sciences, as a way to better understand human thinking and human behaviours (Onwuegbuzie & Leech, 2004). Pragmatist researchers are comfortable in using multiple methodologies or mixed-method approaches to answer their research questions.

In recent decades, there has been an increasing and legitimised 'use of multiple approaches in answering research questions, rather than restricting or constraining researchers' choices' (Johnson & Onwuegbuzie, 2005: 17). Contemporary studies are increasingly engaging in multi-paradigmatic research, where the design of research studies is focused on the integration of multiple methodologies and values drawn from two or more paradigms (Taylor & Medina, 2013). Such a research design goes beyond simply applying multi-methods or mixed-method approaches; that is, a post-positivist using qualitative methods in a mostly objectivist study. Instead multi-paradigmatic research leans on multiple paradigms to design new hybrid methodologies that involve multiple epistemologies, ontologies and their underlying value systems (Taylor *et al.*, 2012). Some authors in this tome acknowledge how limiting one single paradigm can be and discuss how they explored the use of multi-paradigmatic approaches to design their research projects.

Conclusion

As you grow your wings as a researcher, decisions around aspects relating to paradigmatic issues will become easier. Needless to say, your own personal training and/or experiences with certain paradigms, research methodologies and methods are likely to influence your choices (Creswell, 2014). The dandelion image on the front cover of this book captures the esoteric nature of the journey of exploring paradigms and the associated philosophical underpinnings. Such a journey can seem abstract, fuzzy, or irrelevant, just like the dandelion which may seem to be a common and insignificant plant with its feathery pappus that takes the seed on its interminable flight.

As with any journey, for some of you, the thought of paradigms can be perplexing and intense, while for others, engagement with the philosophical aspects of research is interesting and stimulating. So too with the dandelion and its complex mechanism for seed dispersal, where watching the seed gliding in the breeze and parachuting to its destination can seem immaterial to some but for others, it captures their imagination and is a fascinating phenomenon.

Regardless of your stance, the exploration of paradigms is necessary for you if you are serious about your HDR candidature and seek affirmation as an emerging scholar. The dandelion's survival is grounded by its ontological being in a way that your research is substantiated by your own ideologies and values. Your paradigm drives your every action as you travel on your respective life and research journey.

In the ultimate triumph, the escaping seeds of the dandelion symbolise you as the HDR candidate maturing on your journey to becoming a scholar and gaining your 'wings'. In a further parallel, as you, the HDR candidate, fly high into the unknown, you come into contact with other 'seeds' or scholars on their own personal journey, and together you may glide, perhaps even soar, through the rhetoric of paradigms, share ideas and thoughts; and all the while, growing in purpose and fervour. Then one day, just like the dandelion seed that lands into fertile ground and emerges as a new dandelion plant, you, the HDR candidate, find your niche and are transformed into an Early Career Scholar.

References

Alexander, H.A. (2006) A view from somewhere: Explaining the paradigms of educational research. *Journal of Philosophy of Education* 40 (2), 205–221.

Carpenter, B.S., Munoz, O., Munoz, M., Arcak, C., Cornelius, A. and Boulanger, B. (2011) Re/searching for clean water: Artists, community workers and engineers in partnership for positive community change. In C. McLean (ed.) *Creative Arts in Research for Community and Cultural Change* (pp. 41–46). Calgary: Detselig/Temeron Press.

Chilisa, B. and Kawulich, B. (2012) Selecting a research approach: Paradigm, methodology and methods. In C. Wagner, B. Kawulich and M. Garner (eds) *Doing Social Research: A Global Context* (pp. 51–61). New York: McGraw-Hill Higher Education.

Creswell, J.W. (2014) The selection of a research approach. In J.W. Creswell (ed.) *Research Design: Qualitative, Quantitative, and Mixed Methods Approaches* (pp. 3–23). Thousand Oaks, CA: Sage.

Cutcliffe, J.R. (2003) Re-considering reflexivity: The case for intellectual entrepreneurship. *Qualitative Health Research* 13 (1), 136–148.

Denzin, N.K. and Lincoln, Y.S. (2005) *The SAGE Handbook of Qualitative Research* (3rd edn). Thousand Oaks, CA: Sage Publications.

Gage, N.L. (1989) The paradigm wars and their aftermath a 'historical' sketch of research on teaching since 1989. *Educational Researcher* 18 (7), 4–10.

Guba, E.G. (1990) The alternative paradigm dialogue. In E. Guba (ed.) *The Paradigm Dialogue* (pp. 17–27). London, Sage.

Guba, E.G. and Lincoln, Y.S. (1994) Competing paradigms in qualitative research. In N.K. Denzin and Y.S. Lincoln (eds) *Handbook of Qualitative Research* (pp. 105–117). Thousand Oaks, CA: Sage.

Guba, E.G. and Lincoln, Y.S. (2005) Paradigmatic controversies, contradictions, and emerging confluences. In N.K. Denzin and Y.S. Lincoln (eds) *The SAGE Handbook of Qualitative Research* (3rd edn) (pp. 191–215). Thousand Oaks, CA: Sage.

Johnson, R.B. and Onwuegbuzie, A.J. (2005) Mixed method research: A research paradigm whose time has come. *Educational Researcher* 33 (7): 14–26.

Killion, L. and Fisher, R. (2018) Ontology, epistemology: Paradigms and parameters for qualitative approaches to tourism research. In W. Hillman and K. Radel (eds) *Qualitative Methods in Tourism Research: Theory and Practice* (pp. 1–28). Bristol: Channel View Publications.

Kuhn, T.S. (1962) *The Structure of Scientific Revolutions*. Chicago: University of Chicago Press.

McGregor, S.L.T. and Murnane, J.A. (2010) Paradigm, methodology and method: Intellectual integrity in consumer scholarship. *International Journal of Consumer Studies* 34, 419–427.

Mertens, D.M. (2005) *Research and Evaluation in Education and Psychology: Integrating Diversity with Quantitative, Qualitative and Mixed Methods* (2nd edn). Thousand Oaks, CA: Sage.

Munar, A.M. and Jamal, T. (2016) What are paradigms for? In A.M. Munar and T. Jamal (eds) *Tourism Research Paradigms: Critical and Emergent Knowledges* (pp. 1–16). Bingley: Emerald Group Publishing.

Oakley, A. (1999) Paradigm wars: Some thoughts on a personal and public trajectory. *Journal International Journal of Social Research Methodology* 2 (3), 247–254.

Onwuegbuzie, A.J. and Leech, N.L. (2004) Enhancing the interpretation of significant findings: The role of mixed methods research. *The Qualitative Report* 9 (4), 770–792. Retrieved 15 December 2019, from http://www.nova.edu/ssss/QR/QR9-4/onwuegbuzie.pdf

Onwuegbuzie, A.J. and Leech, N.L. (2005) Taking the 'Q' out of research: Teaching research methodology courses without the divide between quantitative and qualitative paradigms. *Quality and Quantity* 39, 267–296.

Patton, M.Q. (2002) *Qualitative Research and Evaluation Methods* (3rd edn). Thousand Oaks: Sage.

Robson, C. (2002) *Real World Research: A Resource for Social Scientists and Practitioner–Researchers* (2nd edn). Oxford: Blackwell.

Sousa, V.D., Driessnack, M. and Mendes, I.A.C. (2007) An overview of research designs relevant to nursing: Part 1: Quantitative research designs. *Revista Latino-Americana de Enfermagem* 15 (3), 502–507.

Taylor, P.C. and Medina, M. (2013) Educational research paradigms: From positivism to multiparadigmatic. *Journal for Meaning-Centered Education* 1, 1–16.

Taylor, P.C., Taylor, E. and Luitel, B.C. (2012) Multi-paradigmatic transformative research as/for teacher education: An integral perspective. In B.J. Fraser, K.G. Tobin and C.J. McRobbie (eds) *Second International Handbook of Science Education* (pp. 373–387). Dordrecht: Springer.

Viglia, G. and Dolnicar, S. (2020) A review of experiments in tourism and hospitality. *Annals of Tourism Research* 80, DOI: https://doi.org/10.1016/j.annals.2020.102858.

2 Logical Positivism in Consumer Behaviour Research

Antje R.H. Graul

Introduction: A Causal Relationship

In order to introduce the philosophical idea of logical positivism (Benton & Craib, 2010), it is crucial to first understand what a cause-and-effect relationship signifies. A causal relationship between two entities suggests that the existence of the second entity is enabled solely as a result of the first entities' existence. Specifically, this implies the notion of replicability of the causal relationship with comparable entities. A causal effect is hence assumed to be observable and can be assembled over time as a result of various observations made by the human mind (Davidson, 1967).

From a logical positivism perspective, it is therefore proposed that based on the regularity of such events, it is possible to test cause-and-effect relationships in a quantitative manner. Consequently, following the philosophy of logical positivism, social science researchers believe in the idea that knowledge can be acquired and derived from observations of an external reality (Graul, 2014), wherefore its falsification can be obtained through multiple observations targeted specifically at the intended analysis (Thorpe, 2017) to confirm and extend suggested theories.

However, despite the popularity of the described research tradition, the underlying philosophical assumptions are often implicit. Subsequently, it seems necessary to elucidate the main philosophical positions that form the basis for quantitative research and their contemporary application to gain statistical inference of causes and effects. This involves specific convictions about first ontology, that is the nature of reality; and second epistemology, that is the nature of knowledge. Next, I aim to reiterate the historical and philosophical tenets of the logical positivism paradigm.

Historical and philosophical tenets of the logical positivism paradigm

In the following, I will focus on the illustration of selected historical and philosophical concepts associated with the logical positivism paradigm. To date, a plethora of philosophical theories have investigated the subject of defining cause, effect and causal relationship. The often contradictory views of philosophers and historical debates regarding this matter can be referred to as the causation debate (Clatterbaugh, 1999). To illustrate the tenets of this debate, I start out by presenting the theory of causation brought forward by the British empiricist Hume. Second, I demonstrate the influence of Hume's assumptions on logical positivism. Opposed to this is Popper's criticism and his suggested alternative, commonly referred to as falsificationism, which I will exemplify thereafter. This is further developed by Kuhn who refuses the main assumptions of logical positivism and the Popperian falsificationism. Finally, I conclude by illustrating Lakatos' response to Kuhn, which maintains elements of Popper's philosophy in light of the Kuhnian challenge.

Hume's theory of causation

David Hume, an empiricist of the 18th century, is famous for his description of requirements necessary to conclude that a true relationship between cause and effect exists. Hume defines a causal relationship as 'an object followed by another, and where all the objects, similar to the first, are followed by objects similar to the second. Or, in other words, where if the first object had not been, the second never had existed' (Hume, 1748/1963, Section VII).

Hume's theory can be illustrated with the help of the following example: An individual may observe that a man uses a lighter to set his chimney wood on fire. It is hence a creation of the human mind that the lighter was the cause that resulted in the effect of fire, as the mind 'forms the idea of cause and effect' based on observations (White, 1990: 4). The chimney wood would not have been set on fire without the lighter, so the first objects' existence is pivotal for the second, which is the fire. Further, the observed effect is replicable with similar objects, which means fire can be sparked with similar chimney wood and a similar lighter. However, it is worth considering that underlying powers or structures of the causal relationship are non-observable (Nadler, 1999).

Let me summarise in detail what this means taking ontology and epistemology into consideration: firstly, causes and effects are seen as *objects* secondly, their causal relationship is assumed to be *replicable* with similar objects; and thirdly, the presence of the second object is solely possible due to the first object's *existence*. Consequently, all these premises involve *observable objects* and therefore imply the ontology that causal relations are constructed from experiences observed by the human mind. With reference to epistemology, this approach further emphasises experience as source of knowledge.

Hume's philosophical approach did, although in a very extreme way, mark a beginning of a shift away from metaphysical considerations towards solely epistemological issues within the causation debate in modern philosophy. In terms of replicability, it is important to say that Hume's theory is still influential on present social science research, as for scholars, 'a causal succession is supposed to be a succession that instantiates a regularity' (Lewis, 1973: 556). However, one major point of criticism that emerging scholars in the fields of social sciences need to be aware of refers to the fact that Hume failed to consider either the possible existence of several causes for one effect (Mill, 1967) or the impossibility of verifying a true reality. Hume's basic assumptions were consequently discussed and further developed as illustrated within the next section.

Logical positivism

One further development of Hume's theory can be defined as logical positivism and can be traced back to the Vienna circle – a group of scientists and philosophers in the 1920s, with whom Karl Popper interacted. With regard to ontology, Easterby-Smith (2008: 57) pointed out that Hume's approach implies 'an ontological assumption that reality is external and objective; and second, an epistemological assumption, that knowledge is only of significance if it is based on observations of this external reality'. Consequently, logical positivists stress the importance of inducing scientific theories via empirical and objective observations of the external reality, not merely observations.

However, this can involve a problematic element, as the following case illustrates: One scientist may observe that every ape he comes across within his survey likes bananas. Hence, he concludes that every ape likes bananas. Nonetheless, there might be an ape on another continent who does not like bananas. The prior inductive inference that all apes like bananas would hence be wrong. That is, although the scientist surveyed constant conjunctions by objectively observing the external world, the conclusion he drew was wrong. This problem was originally introduced by Mill (1967) and developed further by the philosopher Popper to illustrate the *problem of induction* as demonstrated in the following.

Popperian falsificationism

In the 20th century, the Austrian-British philosopher Karl Popper took the problem of induction as an initial point to illustrate his criticism on Hume's theory of causation on the stream of logical positivism. This is very important for understanding the tenets of social science research today. Firstly, Popper opposed the empiricist's view on the theory of knowledge and rather stressed the importance of *the reason* itself as a source of information, with the mind being responsible for *the reason*. Further, as seen in the example of apes liking bananas, logical positivism supposes that a verification of a scientific law is possible based on observation. However, as there might be an ape not liking bananas, Popper concludes that scientists can only confute scientific laws, but are never in the position to verify them. That is, hypotheses can be developed based on proposed theories, and falsified by the deductive use of generating knowledge, as shown in the example by the discovery of an ape not liking bananas, but social science research is suggested to evolve around falsification as verification is impossible.

This supports a provisional approach to scientific knowledge, emphasising that the context of justification is fundamental. As Popper hypothesises the existence of an actual reality but concludes that the observation of this reality is often unsatisfactory and only probabilistically apprehendable, his philosophy is oftentimes referred to as post-positivism (Guba & Lincoln, 1994; Suppe, 1977). Although it is commonly accepted that hypotheses can never be confirmed but only conclusively rejected with the help of contradictory proof, Popper's falsificationism can be criticised in terms of its application within scientific communities as illustrated in the following.

Kuhn's critique

Thomas Kuhn replied to Popper by introducing a 'new view' of science in his work *The Structure of Scientific Revolutions* (1962). His main innovation stemmed from the idea to view science as a social process. According to Kuhn, the existence of different scientific communities equally signifies the existence of shared constellations of beliefs within each community. Kuhn's theory elucidates the role of history in science and specific developments within the scientific community. Therefore, Kuhn introduced the term *paradigm* in order to describe 'the progress of scientific discoveries in practice' (Easterby-Smith, 2008: 57). A paradigm includes shared assumptions, values or beliefs that are generally accepted within one scientific community (Gelo, 2012). Kuhn asserted that once a scientific community is convinced of a paradigm, scientists might be less eager to falsify a commonly accepted paradigm and follow it over centuries.

Furthermore, Kuhn put forward that a new and an old paradigm can involve major conceptual differences. This can be seen in the following example: Under the assumption that the earth is one planet of a planetary system surrounding the sun, the sun rises because the earth rotates. However, under the assumption that the earth is a plate and the centre of the planetary system itself, the sun rises because the sun orbits the earth. This conceptual difference among the paradigms involves a huge potential for misunderstanding between two scientists of which one believes in the first paradigm and the other in the second paradigm. Subsequently, everything that a social scientist observes is assumed to be to a certain degree influenced by prior knowledge and assumptions (Franklin *et al.*, 1989). Kuhn defined this enigma as *the theory of ladenness of observation.*

Second, Kuhn claimed that it is impossible to falsify a scientific law in isolation, as there are always auxiliary assumptions involved and a scientist cannot specify all of them. Referring back to the method of testing a target hypothesis via falsification, the testing of one hypothesis is not possible, as a target hypothesis must always be tested in conjunction with its auxiliary hypotheses. This conundrum is known as *the Duham-Quine thesis* and suggests that a hypothesis cannot be falsified in isolation (Shadish *et al.*, 2002).

To summarise, the incongruity of Kuhn's theses with both the basic assumptions of the stream of logical positivism and Popper's falsificationism becomes apparent. Whereas other philosophers implied that hypotheses can be falsified in isolation, Kuhn contradicts this assumption. Moreover, Kuhn's colleagues presupposed that every causal claim that is operationalised can be correctly defined and understood by different scientists; suppositions which are rejected by Kuhn's 'theory of ladenness of observation'.

Lakatos' response

Based on the method of falsification proposed by Popper, the Hungarian philosopher Imre Lakatos sought to develop a more sophisticated version of falsificationism. Taking Kuhn's critique into consideration and addressing the weaknesses discovered, Lakatos aimed to consider a whole system of hypotheses rather than individual scientific laws. This can be seen as his response to Kuhn's idea of theory ladenness of observations (Blaug, 1975). Therefore, he redefined the term *paradigm* and labelled it as '*scientific research program*' (Lakatos, 1978) and emphasised the existence of a programme that involves auxiliary assumptions and consists of different elements. These elements were summarised by Lakatos (1978) as follows: The core represents specific unquestionable concepts, which are in the centre of scientific research communities and highly unlikely to change. He contended that the surrounding area symbolises different hypotheses that evolve around the core principle and describe further causal relationships based on such. These hypotheses are more likely to be revised or falsified by scientists and can be subject to further testing and empirical analysis (Lakatos, 1978).

These characteristics are very similar to Kuhn's ideas, with the only difference being that Lakatos considers a series of theories and defines two objectives that must be fulfilled in order to achieve progressive development for social science researchers. The first objective describes a progress on a conceptual level that is conceptualising a new scientific law; the second requires a progress on an empirical level, which provides statistical evidence that the new hypothesis cannot be falsified and can therefore be provisionally retained. As both of the two conditions, conceptual and empirical, are fulfilled, the prior theory should be replaced by the new one. This implies a process of development within the scientific community rather than the complete rejection of prior assumptions as suggested by Kuhn.

In the next section, I will build on this extensive review of the historical background that led to the logical positivism paradigm and introduce its application to current research projects. Specifically, the application of experimental designs to manipulate causal or independent variables in a controlled setting in order to investigate their effect on the dependent variable of interest is a common method used by social science researchers today.

Application of the Logical Positivism Paradigm in Social Science Research

Within social science research, these fundamental concepts of falsification are often employed when hypothesised cause-and-effect relationships are investigated in a quantitative manner. This practice is particularly applicable to my own area of research – the consumer behaviour space – and indeed is of continued relevance to my research projects.

For example, in one project, a colleague and I were planning to investigate the potential effect of scarcity cues on booking intentions on hospitality platforms. Specifically, we wanted to find out which marketing cue was most effective in driving consumer booking intentions online – a supply-based scarcity cue (e.g. 3% left for these dates) or a popularity-based scarcity cue (e.g. 3000 other people are currently looking for a place to stay) (Teubner & Graul, 2020). Opposing predictions can be made based on previous research with regard to what scarcity cue may be most effective in driving consumer booking intentions. Thus, it was important to conceptualise a quasi-experimental design that would allow us to test the distinct and combined effectiveness of these cues.

Oftentimes, emerging scholars' research aims similarly lie in investigating the relationship between two independent variables and one dependent variable; which is generally referred to as the 'outcome' variable; in my particular case, booking intentions would be the outcome variable. Hypotheses commonly represent the starting point for such investigations. With regard to my own research project, my colleague and I developed a set of hypotheses based on previous findings, relevant literature and the resulting theorising. Specifically, we proposed the following (Teubner & Graul, 2000: 4):

> Hypothesis 1 (The Get-It-before-It's-Gone Hypothesis). Perceived scarcity is a) positively associated with perceived urgency which b), in turn, is positively related to booking intentions.

Hypothesis 2 (The Must-Be-Good Hypothesis). Perceived scarcity is a) positively associated with perceived value which b), in turn, is positively related to booking intentions.

In order to examine the relationships proposed in these two hypotheses, we chose to follow a quantitative approach that allowed us to investigate the differential effects of the independent variables (supply-based and popularity-based scarcity cues) on the dependent variable (consumer booking intentions) with the help of collected survey data analysed through statistical analysis software. This methodology reflects the idea of a logical positivism philosophy (Benton & Craib, 2010) which assumes the regularity of events and suggests that cause-and-effect relationships can be tested in a quantitative manner and 'analysed in numerical form' (Gelo, 2012: 113). This emphasises the logical positivists' stand on the importance of developing scientific theories via the sum of hypothesising, objective observations and empirical testing of the external reality.

Causality in experimental designs

In line with logical positivism philosophy, implementing an experimental research design similar to the example provided above is oftentimes the most appropriate method for me when it comes to investigating consumer behaviour and decision-making. This approach allows me to investigate my proposed hypotheses and test the related causal claims (Cozby & Bates, 2012). Within social science research, the quasi-experimental design has proven its suitability for testing causal relationships in a controlled setting, in which 'the causal or independent variables are manipulated in a relatively controlled environment' (Malhotra *et al.*, 2012: 108).

An experimental design describes a well-established method in which, based on a factor with different levels, stimuli for each level can be created and tested under identical conditions in a relatively controlled environment (Malhotra *et al.*, 2012). As a result, a change in the dependent variable will be attributed to a change in the levels of the independent factors implied. Thus, particularly with regard to designing the stimuli material and setting up the research design, it was crucial for me to follow the order of occurrence correctly, involving an exposure to the stimuli followed by a measurement of the dependent variable. In fact, in order for a causal relationship to exist, three conditions need to be fulfilled: the 'concomitant variation, time order of occurrence of variables, and absence of other possible causal factors' (Malhotra *et al.*, 2012: 252). By following these conditions, the causal effect of independent variables on dependent variables can be analysed with the help of empirical data (Gelo, 2012) that are collected after the exposure to the experimental stimuli, commonly involving self-report measures of the operationalised factors within a self-reporting survey (Graul, 2014).

In order to test for causality, it is important to remember the method of falsification in order to evaluate whether the empirical data provide sufficient evidence to provisionally support or reject the null hypothesis (Graul, 2014). As introduced earlier in this chapter, following the philosophical ideas of Popper and Lakatos (Benton & Craib, 2010), there is a need to emphasise that, in case of a provisional rejection of the null hypothesis, the opposed claim that suggests that a causal relationship between the two variables exists can be provisionally accepted. Nonetheless, while falsification can be conducted, an omnifarious validation of the hypothesis as a result of an observed phenomenon would be impossible to obtain due to the potential for flawed and erroneous observations by the human mind (Benton & Craib, 2010). Hence, I was mindful that in line with logical positivism, further empirical testing would need to be conducted to seek support for the provisionally accepted hypotheses.

Internal and external validity in experimental designs

Next, it is important to consider matters of validity. We can distinguish between two different forms of validity: (1) internal validity and (2) external validity. First, internal validity describes the degree to which the experimental stimuli can be identified as the responsible cause for the changes observed in the dependent variable within each experimental condition and is usually assumed to be high within experimental designs due to the high level of environmental control (Campbell & Stanley, 2015). A high level of internal validity is therefore a crucial precondition for every experimental design. In particular, this reduces the complexity of real-life settings in favour of the intended investigation. As a result, a setting in which 'variables are manipulated and their effects upon other variables observed' can be created (Campbell & Stanley, 2015: 1). A highly controlled setting entails high levels of consistency throughout the study and the potential to eliminate and/or control for plausible confounds and environmental factors influencing results (Monette *et al.*, 2005). Thus, it is absolute imperative to spend sufficient time elaborating the experimental setup and conducting all experiments following 'well designed, carefully controlled, and meticulously measured' characteristics (Druckman *et al.*, 2011: 28).

Second, external validity describes the degree to which the experimental findings are generalisable to other populations, segments or measurement variables (Campbell & Stanley, 2015) and replicable in various contexts. As such, laboratory experiments can be criticised for their lack of transferrable results that are applicable to real-life situations. In support of this criticism, researchers have long expressed apprehension that 'there exists a concern that much of consumer research, and behavioural research in general, is not generalisable' due to the gathering of artificial data (Calder *et al.*, 1981: 197). This problem is oftentimes addressed by involving additional field experiments or analysis of secondary data in the marketplace.

Coming back to the project introduced above, it was crucial for my colleague and I to examine the effect of the distinct scarcity cues as independent factors in a relatively controlled environment in which changes in the outcome variable (consumer booking intentions) can be attributed to changes in the experimental stimuli with a high likelihood. I found this particularly relevant to my own research and so explored it further so as to better understand this approach. My endeavours led me to the specific characteristics of two popular experimental designs in social science research, namely the *one-way, factorial* and *between versus within-subjects* designs, which are reviewed in the following.

One-way and factorial designs

While a one-way experimental design is concerned with the main effect of changes in one independent variable on the outcome variable, this only involves one independent variable and its respective levels; a factorial design permits changes of diverse factors, thus two or more independent variables with their different levels and respective interplay, to be taken into consideration. Consequently, within the scope of the previously introduced social science research approach that aimed to investigate the effect of two cues (two factors), I would recommend conducting a two-way design in order to investigate direct and interaction effects of two independent variables on one outcome variable of consumer booking intentions (2 x 2). As a result, four experimental groups emerge, as illustrated below using my own research (see Table 2.1).

Thus, following the approach outlined in Table 2.1, my colleague and I were able to test four different scenarios: one that showed a popularity-based cue, one that showed a supply-based cue, one that showed a combination of both a popularity- and supply-based cue and one that did not show any cue. After exposing study participants to one of the four conditions, we measured potential outcome variables such as consumer booking intentions and were able to identify potential differences in booking intentions among those four groups (Teubner & Graul, 2020). While for the presented research project, a between-subjects design was chosen, I would like to elaborate more on the designs that can be used for similar research projects in the following.

Table 2.1 Factorial design

		Factor 1: Supply-Based Cue	
		Level 1: Present	Level 2: Absent
Factor 2: Popularity-Based Cue	Level 1: Present	Group A	Group B
	Level 2: Absent	Group C	Group D

Between- and within-subjects designs

In order to collect data in a way that is most appropriate for the proposed factorial design, I oftentimes decide to employ between-subjects experimental designs for my intended data collections based on the following three reasons. First, while a within-participant design exposes all recruited respondents to each of the designed experimental stimuli, a between-participants design allocates different respondents to one of the experimental groups solely. As a result, within-participant designs bear a higher risk of jeopardising the independence of the exposure to diverse stimuli and as a result a risk of leading to erroneous conclusions regarding the causal estimates (Charness *et al.*, 2012). Second, research suggests that in a simplified way, demand effects are expected to be higher when incorporating within-participants designs, as a result of a pattern that participants may aim to follow based on their envisioned research objective of the experiment (White, 1977). For example, if one participant in our study had seen all four groups, it may be likely that the participant draws his own conclusions regarding our proposed hypothesis and while his first response may have been accurate, his second, third and fourth response may be influenced by the responses the participant had previously given. Third, the between-subjects design has been proven to be particularly accurate for investigating problems or choices that are close to the consumers' behaviour in the marketplace (Charness *et al.*, 2012), such as consumer booking intentions on a hospitality platform.

Data collection strategy

There are many different ways that allow social science researchers to implement experimental research and collect respective data. For instance, I designed my experimental stimuli involving visual and textual components that can easily be incorporated within an online survey format, and the constructs of interest were then surveyed involving a carefully designed questionnaire.

If this procedure is possible, one efficient way to recruit respondents in consumer behaviour research is to recruit participants online via Amazon's Mechanical Turk (AMT), involving a non-probability and self-selecting sample (Malhotra *et al.*, 2012). AMT is a source of data collection that emerged with the rapid development of online technology and the internet. AMT describes an online platform that brings *requesters* that are looking for respondents to complete their tasks (e.g. a survey) and *workers* that are interested in completing digital tasks (e.g. responding to a survey or writing task) together. Thus, AMT allows scholars to publish their survey programmed with an external survey tool such as Qualtrics on the platform in order to recruit respondents in a rapid and inexpensive way as opposed to the previous reliance on laboratory studies and student samples (Buhrmester *et al.*, 2011). Consequently, a significant growth in publications that rely on AMT samples has been observed recently, with over 400 publications allocated within the field of social sciences (Paolacci & Chandler, 2014).

Following exposure to the experimental stimuli, the data collection is usually intended to involve measures of the constructs of interest by the employment of close ended, 7-point Likert scales and established item batteries that 'require the participants to indicate a degree of agreement or disagreement' (Malhotra *et al.*, 2012: 213). Then, data can be analysed with the help of statistical software such as MS Excel or SPSS.

Reflections and Elaboration of Experimental Designs

In this section, I will conclude with an overview of the resulting advantages and limitations of experimental research that emerging scholars should carefully consider when designing their experimental design.

While a controlled experimental setting similar to the scenario presented above in which we presented hypothetical booking scenarios (Teubner & Graul, 2020) implies ample advantages, limitations emerge due to the artificial nature of the collected data. Thus, internal validity may jeopardise the desired external validity, and obtained results may not be applicable to real-life scenarios. As introduced earlier, the problem of generalisability is the first point that needs to be addressed. While Popper proposed that, based on observation solely, a verification of scientific laws is impossible, this suggests that results from observation cannot be universally generalised and therefore only provisionally accepted. In academic research, scholars aim to reach a high level of internal and external validity to improve this issue. As reviewed prior to this paragraph, internal validity describes the extent to which the test measures what it aims to survey (Monette *et al.*, 2005), while external validity 'refers to the generalisability of findings from a study' (Druckman *et al.*, 2011: 34). A carefully developed experimental design hence tries to control complexity by eliminating confounding effects within a constant laboratory setting to prove a causal relationship. However, it can be criticised for its difficulty to transfer results into real-life settings. That is, the generalisability of laboratory experiments is strongly limited due to the collection of artificial data. In contrast to laboratory settings, field experiments offer an alternative in which an experiment might be piloted 'under actual market conditions in a real-life-setting' (Malhotra, 2012: 272). This, however, involves the risk that field experiments are influenced by confounding variables within the real environment. Consequently, a critical matter of discussion is the impossibility to entirely control an experimental setting (Lewis, 1973: 558). In addition to conducting combinations of experimental data collections that employ both high internal and high external validity, scholars commonly make use of mathematical testing to investigate the reliability of an experimental design. The reliability test involving Cronbach's Alpha is oftentimes used within social science research and designed to test the level of internal consistency.

Further, it is worth noting that despite a high level of control, distinct variables unknown to the researcher, such as health-related issues and personal biases, may not be controllable, and may affect the results obtained in a particular way. Therefore, unreliable samples are potential risks when conducting experiments.

Consequently, it is important for emerging scholars to be aware of the fact that internal validity may in numerous cases jeopardise external validity of experimental research, and vice versa (Campbell & Stanley, 2015). This challenge is often addressed by involving a combination of different data collection strategies (such as online experiment, laboratory experiment, field experiment) and various forms of manipulations of the independent variables in order to demonstrate the effects' robustness within different settings. For example, in our research project, we conducted an additional study in the field to introduce our quasi-experimental design. Specifically, we incorporated a web-crawler to collect data on the prevalence of scarcity cues in real life from two leading hospitality platforms. During this field procedure, we identified that supply-based scarcity cues were shown in 58.9% of the search queries submitted by the web-crawler, while popularity-based cues were shown in 48.5% of all cases (Teubner & Graul, 2020). This helped us to provide a first account and identify the prevalence of our factors in real-life scenarios.

From a personal point of view, my experience has shown that reviewers may criticise manuscripts if those rely too heavily on one particular data collection strategy that does not provide high levels of external validity. Thus, I would recommend that emerging scholars try to involve a combination of approaches. Despite acknowledging limitations of experimental research designs, I would recommend to emerging scholars that following a quasi-experimental design to examine proposed causal hypotheses is most suitable in most consumer behaviour and decision-making research. Together, this allows emerging scholars to investigate their research questions from the perspective of a logical positivism paradigm, bringing validity and rigour to their social science research.

References

Benton, T. and Craib, I. (2010) *Philosophy of Social Science: The Philosophical Foundations of Social Thought*. Basingstoke: Palgrave Macmillan.

Blaug, M. (1975) Kuhn versus Lakatos, or paradigms versus research programmes in the history of economics. *History of Political Economy* 7 (4), 399–433.

Buhrmester, M., Kwang, T. and Gosling, S.D. (2011) Amazon's Mechanical Turk: A new source of inexpensive, yet high-quality, data? *Perspectives on Psychological Science* 6 (1), 3–5.

Calder, B.J., Phillips, L.W. and Tybout, A.M. (1981) Designing research for application. *Journal of Consumer Research* 8 (2), 197–207.

Campbell, D.T. and Stanley, J.C. (2015) *Experimental and Quasi-experimental Designs for Research*. Cambridge: Ravenio Books.

Charness, G., Gneezy, U. and Kuhn, M.A. (2012) Experimental methods: Between-subject and within-subject design. *Journal of Economic Behavior and Organization* 81 (1), 1–8.

Clatterbaugh, K. (1999) *The Causation Debate in Modern Philosophy, 1637-1739*. New York: Routledge.

Cozby, P.C. and Bates, S. (2012) Methods in behavioral research. New York, NY: McGraw-Hill.

Davidson, D. (1967) Causal relations. *The Journal of Philosophy* 64 (21), 691–703.

Druckman, J.N., Green, D.P., Kuklinski, J.H. and Lupia, A. (2011) *Cambridge Handbook of Experimental Political Science*. Cambridge: Cambridge University Press.

Easterby-Smith, M. (2008) Chapter 4: The philosophy of management research. In M. Easterby-Smith (ed.) *Management Research* (3rd edn, pp. 55–79). London: SAGE.

Franklin A., Anderson, M., Brock, D., Coleman,S., Downing, J., Gruvander, A., Lilly, J., Neal, J., Peterson, D., Price, M., Rice, R., Smith, L., Speirer, S. and Toering, D. (1989) Can a theory-laden observation test the theory? *The British Journal for the Philosophy of Science* 40 (2), 229–231.

Gelo, O. (2012) On research methods and their philosophical assumptions. *Psychotherapie and Sozialwissenschaft* 2, 109–128.

Graul, A. (2014) *The Causation Debate: A philosophical perspective on experimental design in consumer behaviour research*. LUBS5287M, University of Leeds.

Guba, E.G. and Lincoln, Y.S. (1994) Competing paradigms in qualitative research. In N.K. Denzin and Y.S. Lincoln (eds) *Handbook of Qualitative Research* (pp. 105–117). Thousand Oaks: Sage Publications.

Hume, D. (1963) *An Enquiry Concerning Human Understanding*. La Salle: Open Court Press (original work published 1748).

Kuhn, S.T. (1962) *The Structure of Scientific Revolutions*. Chicago: University of Chicago Press.

Lakatos, I. (1978) *The Methodology of Scientific Research Programmes*. Cambridge: Cambridge University Press.

Lewis, D. (1974) Causation. *The Journal of Philosophy* 70 (17), 556–567.

Malhotra, N.K., Birks, D.F. and Wills, P.A. (2012) *Marketing Research. An Applied Approach* (4th edn). Harlow: FT Prentice Hall.

Mill, J.S. (1967) *A System of Logic, Ratiocinative and Inductive*. London: Longmans. (Original work published 1843).

Monette, D.R., Sullivan, T.J. and DeJong, C.R. (2005) *Applied Social Research. A Tool for the Human Services* (6th edn). Toronto: Thomson Learning Inc.

Nadler, S. (1999) The causation debate in modern philosophy, 1637-1739 by Kenneth Clatterbaugh. *The British Journal of Philosophy of Science* 50 (3), 501–504.

Paolacci, G. and Chandler, J. (2014) Inside the Turk: Understanding Mechanical Turk as a participant pool. *Current Directions in Psychological Science* 23 (3), 184–188.

Shadish, W.R., Cook, T.D and Campbell. D.T. (2002) *Experimental and Quasi-experimental Designs For Generalized Causal Inference*. Boston, MA: Houghton Mifflin.

Suppe, F. (1977) *The Structure of Scientific Theories* (2nd edn). Champaign-Urbana, IL: Illini Books.

Teubner, T. and Graul, A. (2020) Only one room left! How scarcity cues affect booking intentions on hospitality platforms. *Electronic Commerce Research and Applications* 39, 1–11.

Thorpe, R. (2017) Philosophy of Social Science Research. Presented at Q-Step Programme, University of Leeds, Leeds, United Kingdom. July 2017.

White, P.A. (1990) Ideas about causation in philosophy and psychology. *Psychological Bulletin* 108 (1), 3–18.

White, R.A. (1977) The influence of experimenter motivation, attitudes, and methods of handling subjects on Psi test results. *Handbook of Parapsychology* 273–301.

3 The Design Science Research Paradigm: An Instantiation of Website Benchmarking

Leonie Cassidy

Introduction

In this digital age, business problems cannot always be investigated and solved using traditional research methods as advances and changes in technology are occurring at an ever-increasing speed. Therefore, an additional approach is required. Why should we, in the business discipline, only consider our traditional paradigms? Why not consider paradigms available and accepted in other disciplines? After all, most research can now be termed 'cross-disciplinary' as technology is used increasingly to assist in, and at times, is the outcome of research, regardless of the research discipline. Therefore, why not look to the technology disciplines for a paradigm?

Looking to the information systems (IS) discipline we find the design science research (DSR) paradigm. This paradigm has long been situated in the realms of engineering and the sciences of the artificial (Hevner et al., 2004; Iivari, 2007). However, over the last twenty years the DSR paradigm has become increasingly accepted as central to IS research (Hevner et al., 2019). As digital innovation brings rapid changes to business and management, the DSR paradigm becomes more and more relevant to this discipline (Turetken et al., 2019). To understand how the DSR paradigm can be applied to other disciplines I first look to its history, its appearance and use in the engineering discipline, and its path to IS acceptance. I then discuss the DSR paradigm and methodology in relation to my thesis research into website benchmarking, how I applied it to develop a new website benchmarking approach and how I used it in the process of creating my thesis. I conclude with some advice that I hope is useful to the readers of this chapter.

The DSR Paradigm

The science of design had begun to emerge (by 1981) and penetrate computer science, systems engineering and even some business schools offering management science (Simon, 1981). In his well-cited book, *The Sciences of the Artificial,* Simon (1981) describes the difference between the natural sciences and design science as:

> The natural sciences are concerned with how things are...
>
> Design, on the other hand, is concerned with how things ought to be, with devising artefacts to attain goals. (1981: 114)

Although being recognised by 1981, the DSR paradigm still had a long difficult journey towards acceptance. Early on there was a lack of agreement on where the DSR paradigm actually fitted. There were arguments for a broader philosophical base than just positivism and interpretivism; debates on whether design theories were needed, or if DSR was in fact a new paradigm (Hovorka & Germonprez, 2010). Even in the early 1990s it was recognised that there were differences between the DSR paradigm and paradigms used to build and test theories, and interpretative research (Peffers *et al.*, 2007). However, this all began to change with the publication of the seminal paper by Hevner *et al.*, in 2004. In the paper, they provided the following definition of the difference between the behavioural science paradigm and the DSR paradigm:

> ....behavioural science paradigm seeks to develop and verify theories that explain or predict human or organisational behaviour, the design science paradigm seeks to extend the boundaries of human and organisational capabilities by creating new and innovative artefacts. (Hevner *et al.*, 2004: 75)

In particular, the focus was on IS research; they specified that information technology (IT) artefacts are broadly defined as constructs (vocabulary and symbols), models (abstractions and representations), methods (algorithms and practices) and instantiations (implemented and prototype systems) (Hevner *et al.*, 2004). These specifics became an integral part of the DSR paradigm. Hevner *et al.* chose to provide a description of the difference between 'routine design' and DSR:

> Routine design is the application of existing knowledge to organisational problems, such as constructing a financial or marketing information system using best practice artefacts.....design science research addresses important unsolved problems in unique or innovative ways or solved wicked problems in more effective or efficient ways. (2004: 6)

The DSR paradigm requires knowledge and understanding of the problem, the solution and how to construct and apply the artefact. Hence, the following seven guidelines provide direction (adapted from Hevner *et al.*, 2004):

(1) Design as an artefact: a viable artefact (construct, model, method or instantiation) must be produced from the research.
(2) Problem relevance: '...develop technology-based solutions to important and relevant business problems'.
(3) Design evaluation: evaluation methods must be rigorous.
(4) Research contributions: contributions in the 'design artefact, design foundations, and/or design methodologies' must be 'clear and verifiable'.
(5) Research rigor: rigorous methods must be applied in the evaluation and construction of the design artefact.
(6) Design as a search process: 'the search for an effective artefact requires utilising available means to reach desired ends while satisfying laws in the problem environment'.
(7) Communication of research: research and results must be presented in formats acceptable to technology-oriented audiences and management-oriented audiences.

Therefore, the DSR paradigm, requires a designer/researcher to pursue answers to questions relevant to a human/business problem(s) by creating innovative artefact(s), thereby contributing new knowledge to the field. The artefact(s) should be useful and fundamental in understanding the identified problem (Hevner & Chatterjee, 2010).

The philosophical views of the design science researcher may alter as they move through the phases of the ontology, epistemology, methodology and axiology of the design science paradigm (Vaishnavi & Kuechler, 2004). The associated ontological assumption is 'there are multiple, contextually situated alternative world-states, and these are socio technologically enabled' (Vaishnavi & Kuechler, 2004: 20), using an inductive process, and becoming deductive as the research develops. Epistemologically, the researcher explores the nature of knowledge, enabling the creation and testing of an artefact, and producing factual results. Design science is a developmental methodology where impacts of the artefact(s) on the combined system are measured. The axiology delivers value through control, and the creation of new artefact(s) (Vaishnavi & Kuechler, 2004). This results in the delivery of progress, improvement and understanding to the body of knowledge, via communication of results to appropriate audiences (Kuechler & Vaishnavi, 2011).

Peffers *et al.* (2007) were concerned that although the DSR paradigm had been defined, the uptake in IS research had been slow and limited. This may have been, in part, due to the lack of a standardised model, which allows researchers to develop a 'mental model'. Mental models, in research, are all slightly different, and being able to apply a mental model in IS research allows others to understand and evaluate IS publications (Peffers *et al.*, 2007). To this end, Peffers *et al.* (2007) proposed a design science research methodology (DSRM) with six activities based around the seven guidelines discussed by Hevner *et al.* (2004) – as shown in Figure 3.1.

The first activity in the DSRM is 'Problem identification and motivation'. The research problem is first identified which provides direction in the development of the artefact (solution). Here an explanation for the motivation to conduct the research and develop a solution is also discussed. The second activity is 'Define the objectives for a solution'. Here the researcher discusses the objectives for providing a solution to the problem. What is already in existence in relation to the problem, why these solutions are not adequate and why the researcher requires the proposed artefact. The third activity 'Design and development' is when the artefact is created; these artefacts can be 'constructs, models, methods, or instantiations' (Peffers *et al.*, 2007: 49). This activity includes theory, understanding the requirements of what the artefact is required to do, refining the design and creating the artefact. Activity 4 is the 'Demonstration' of the artefact. This is where the artefact is tested to see if it solves the original problem. Following on from the demonstration is the fifth activity, 'Evaluation' where the results expected when the artefact was applied to the problem are compared to what actually occurred. The final activity, the sixth, is 'Communication'; this is where everything is communicated in the appropriate format to the relevant audience(s).

The DSRM is a sequential process. However, the researcher(s) can stop at any step and move back to a previous step (Figure 3.1) to refine, adjust (feedback loops) or even reconsider the artefact for the solution. Therefore, even though DSRM is sequential it is also iterative (Peffers *et al.*, 2007). This iterative capability means DSRM can be a rapid development or solution process. Another advantage of this process is that it permits multiple stages to be developed concurrently.

Since the DSRM was specified by Peffers *et al.* in 2007, acceptance of the use of the DSR paradigm and the DSRM continues to increase. For example, DSRM has been applied in business intelligence systems (BIS) research in Taiwan, with the aim to develop a hospital BIS (Kao *et al.*, 2016). In education research DSRM is being used to discover a better way to 'Gamify Education' (El-Masri & Tarhini, 2015), and in retail to try and develop a service system that permits co-creation of a digital customer experience in a bricks-and-mortar store (Betzing *et al.*, 2018). The following section describes my application of the design science paradigm and DSRM in my PhD research.

Application of the Design Science Paradigm

Traditional higher-degree research theses may consist of sections in the form of chapters; for example, introduction, a series of chapters that includes the literature review, the methods section and a general discussion or conclusion section. However, after completing my initial literature

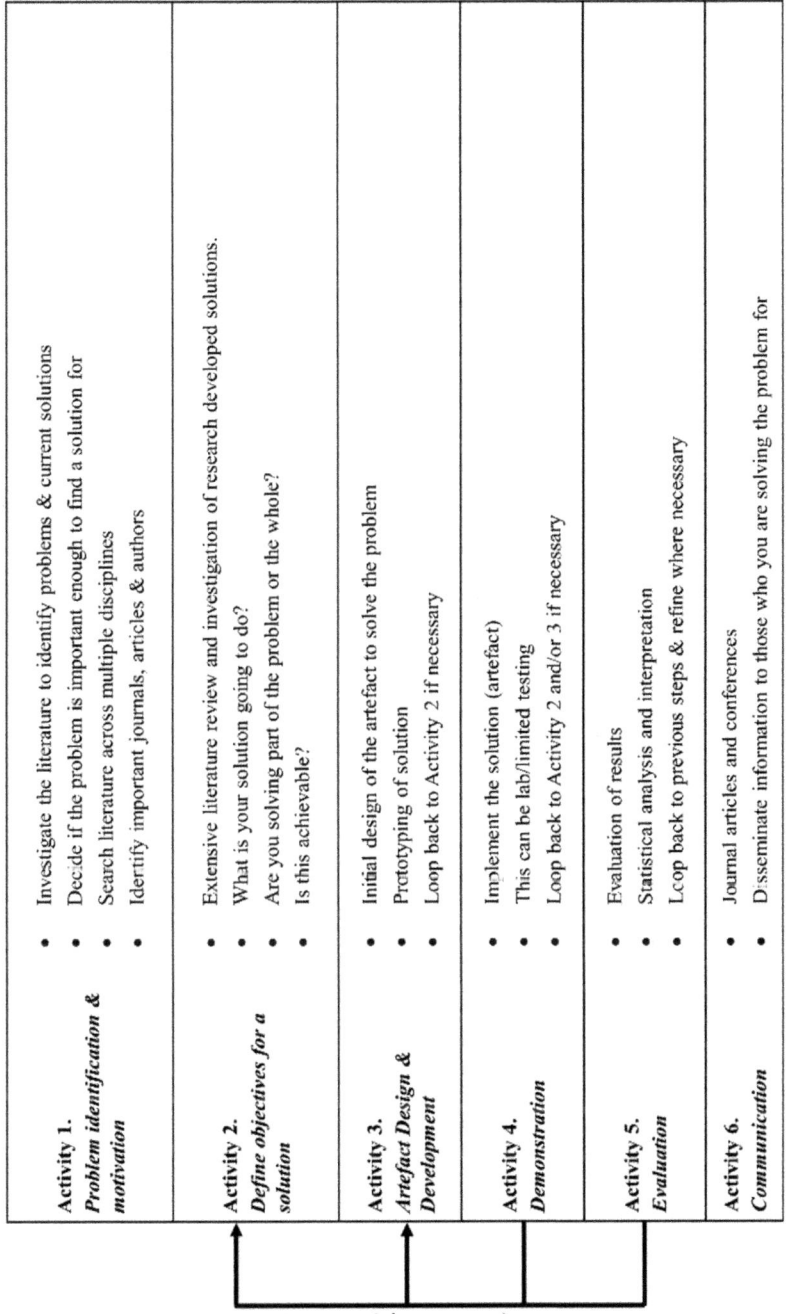

Figure 3.1 Design science research (DSR) staging of artefact solution (Source: adapted from Cassidy & Hamilton, 2016a)

review it became apparent that this approach was not suitable as I was conducting research in the digital realm where everything moves at an increasingly fast pace, and whatever I found would have to be published along the journey instead of at the end.

I struggled to move forward as the methodology I needed for my research was proving elusive. I had investigated the numerous approaches in the literature; however, I really did not want to go down the subjective path as this had been done before in numerous targeted studies. Many of these studies used university students as their respondents, thus introducing more limitations into the results.

I, along with my supervisors, spent many hours with a whiteboard attempting to determine the best way to move forward with my research. Then one day, we are not sure who discovered the article first, myself or my junior supervisor, we both arrived at the senior supervisor's office with the same journal article. It was, the now-seminal article, by Peffers *et al.* (2007), where they make concrete the methodology for the design science paradigm.

Peffers *et al.* (2007) utilised Hevner *et al.*'s (2004) seven guidelines for design science research in information systems, and developed the methodology termed DSRM using six activities, along with a framework. After reading the article we all came to the same conclusion – even though this paradigm was for information systems, it was compatible with my cross-disciplinary research. The paradigm, ontology, epistemology, axiology and now the methodology all synced with what I was attempting to achieve. Hence my research methodology problem was finally solved.

I had identified a business problem, and the motivation to provide a solution (Figure 3.2, activity 1). I have long been interested in e-business, information systems and information technology research, even though I am situated in the business management realm. From the literature I identified a problem for business worldwide – that website benchmarking is time-consuming, poorly measured, generally point-in-time tracked and often expensive, and such studies typically involved manually assessed 'expert' comparisons. Previous approaches to website benchmarking (including those online) were generally survey based, and either focused in-depth in just one specific section of a website or focused superficially in assessing at a general level across a broad area. Even then, no agreement on how, and/or what, to measure on a website had been established. In addition, there was no accepted development of a website benchmarking theory (the motivation).

Figure 3.2, activity 2 collated the extensive literature across disciplines, and determined the current approaches available to solve the business problem, in this case the literature covered benchmarking, websites, business, behaviour, technologies approaches and measurement. Once I established there was no suitable approach that thoroughly benchmarked

The Design Science Research Paradigm   31

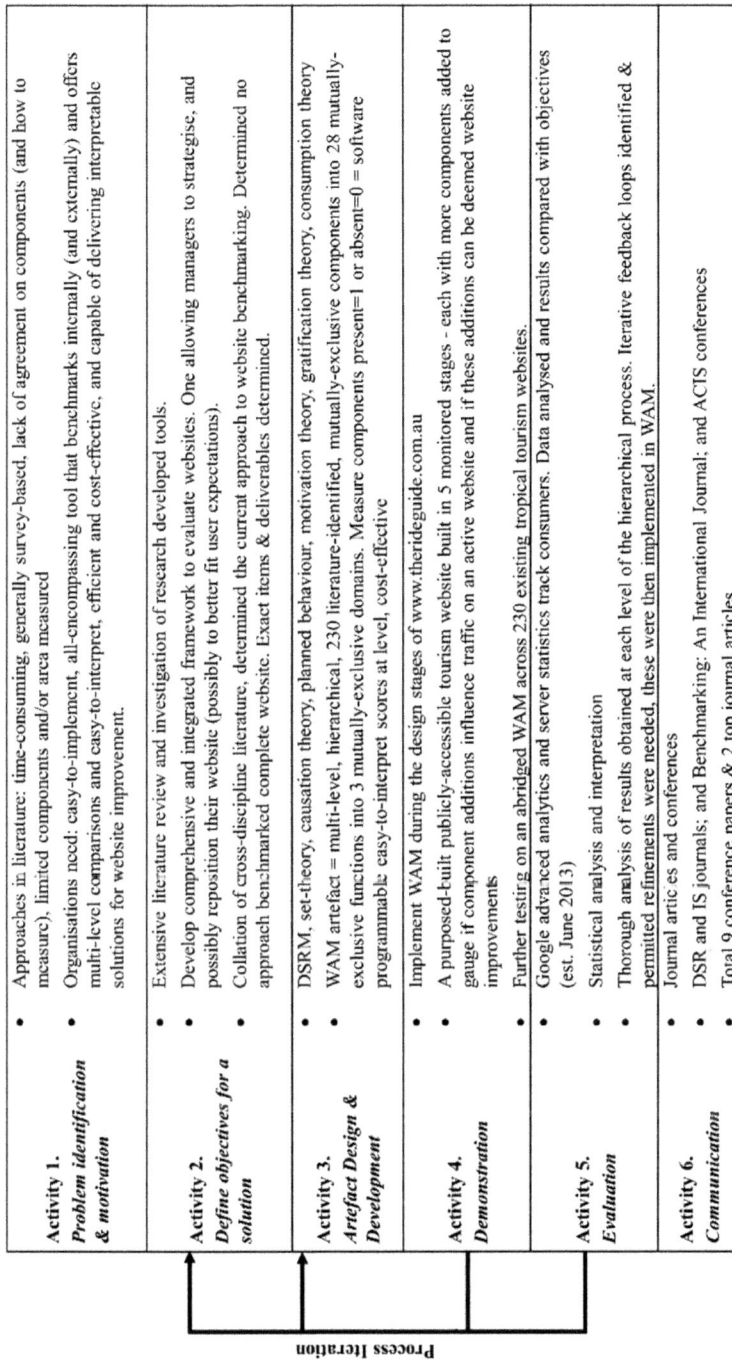

Activity 1. Problem identification & motivation	•	Approaches in literature: time-consuming, generally survey-based, lack of agreement on components (and how to measure), limited components and/or area measured
	•	Organisations need: easy-to-implement, all-encompassing tool that benchmarks internally (and externally) and offers multi-level comparisons and easy-to-interpret, efficient and cost-effective, and capable of delivering interpretable solutions for website improvement.
Activity 2. Define objectives for a solution	•	Extensive literature review and investigation of research developed tools.
	•	Develop comprehensive and integrated framework to evaluate websites. One allowing managers to strategise, and possibly reposition their website (possibly to better fit user expectations).
	•	Collation of cross-discipline literature, determined the current approach to website benchmarking. Determined no approach benchmarked complete website. Exact items & deliverables determined.
Activity 3. Artefact Design & Development	•	DSRM, set-theory, causation theory, planned behaviour, motivation theory, gratification theory, consumption theory
	•	WAM artefact = multi-level, hierarchical, 230 literature-identified, mutually-exclusive components into 28 mutually-exclusive functions into 3 mutually-exclusive domains. Measure components present=1 or absent=0 = software programmable easy-to-interpret scores at level, cost-effective
Activity 4. Demonstration	•	Implement WAM during the design stages of www.therideguide.com.au
	•	A purposed-built publicly-accessible tourism website built in 5 monitored stages - each with more components added to gauge if component additions influence traffic on an active website and if these additions can be deemed website improvements
	•	Further testing on an abridged WAM across 230 existing tropical tourism websites.
	•	Google advanced analytics and server statistics track consumers. Data analysed and results compared with objectives (est. June 2013)
Activity 5. Evaluation	•	Statistical analysis and interpretation
	•	Thorough analysis of results obtained at each level of the hierarchical process. Iterative feedback loops identified & permitted refinements were needed, these were then implemented in WAM.
Activity 6. Communication	•	Journal articles and conferences
	•	DSR and IS journals; and Benchmarking: An International Journal; and ACIS conferences
	•	Total 9 conference papers & 2 top journal articles.

Process Iteration

Figure 3.2 Design science research methodology (DSRM) for website analysis method (WAM) (Source: adapted from Cassidy & Hamilton, 2016a)

the complete website, I could move forward. Here more details were next identified concerning the exact items and their deliverable requirements.

Next, the artefact's design and development (Figure 3.2, activity 3) were closely investigated. As I had previously identified, there was a lack of a definitive theory for website benchmarking. This is where (and in-part why) I deployed the DSR paradigm and DSRM. My research needed a numerical 'cause-and-effect' consumptive and behavioural approach. Hence, I employed a combination of set-theory, causation theory, planned behaviour, motivation theory, gratification theory and consumption theory. The exact application and combination of these theories is available in Cassidy and Hamilton (2016a). Once I had completed development of the supporting theory I began work on the artefact – termed the 'website analysis method' (WAM).

The WAM artefact (Figure 3.3) contained a multi-level, hierarchical structure. It contained 230 mutually exclusive literature-supported measurement item components. Each of these literature-supported measurement item components resides within only one of 29 mutually exclusive functions, and each function resides under only one of three mutually exclusive domains. The WAM artefact was specifically designed to be software programmable, and with an easy-to-interpret benchmarking output (Cassidy & Hamilton, 2016b).

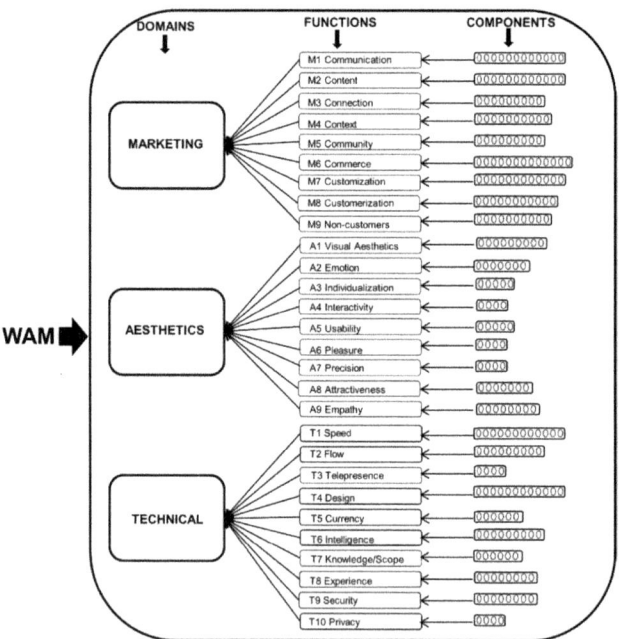

Figure 3.3 The website analysis method (WAM) artefact (Cassidy & Hamilton, 2016a)

Once the artefact was designed and developed, DSRM next required a demonstration (Figure 3.2, activity 4). In this instance WAM was implemented and measured across each of the multi-stages of development of a publicly available tourism website. This tourism website was constructed specifically as a measurement-trial-test-instrument for this website benchmarking research (Cassidy *et al.*, 2014, 2015; Cassidy & Hamilton, 2016a).

Figure 3.2, activity 5 is where evaluation of WAM's performance was measurement item, multi-level analysed down to component level. This application of DSRM also has feedback loops and so an iterative improvement process included exposed adjustments (Peffers *et al.*, 2007) – which then became part of the artefact's website benchmarking solution.

Figure 3.2, activity 6 delivered the results as a set of information that was communicated to appropriate audiences. The journey of the construction and analysis of the WAM artefact was communicated in nine conference papers, and two top journal articles (Cassidy & Hamilton, 2011a, 2011b, 2012a, 2012b, 2013, 2014a, 2014b, 2016a, 2016b; Cassidy *et al.*, 2014, 2015).

The DSR paradigm and DSRM process were both used in the construction of my PhD thesis. The great benefit of this DSRM approach was that you publish continually and generate conference papers, plus journal articles, that fit within sectors of your ongoing and developing research agenda. This also means that, at the same time, you can publish multiple papers separately across any of the first five steps of both the Figure 3.2 and Figure 3.4 DSRM activities. Hence, the sequence of publications may at first appear to be scatter-gun, and not always specifically and/or stepwise staged, but each paper adds in some way to the artefact and its solution. Further, in respect of journal article publications – this can be both a tedious and lengthy process, and consequently these publications may sometimes be completed after a major research thesis or PhD is concluded.

Figure 3.4's sixth DSRM activity step ultimately became the six chapters of my PhD thesis. Each of the published conference papers and journal articles were individually and specifically located, sequenced and positioned within their relevant activity chapter.

Following this process, I could then determine how to relate and link each paper into its relevant PhD thesis chapter. I could then ensure how the combination of published conference papers and journal articles within my PhD thesis flowed together to build a compelling story. This story flowed from the problem identification and motivation and moved through to the concluding communication of my PhD research via my thesis. This final activity 6 step of my thesis linked my PhD thesis publications to my research findings as: real world, theoretical, experimental and business conclusions.

Final Thoughts

As with other PhD students I had a research topic already in my mind when I approached my potential supervisors. They were enthusiastic

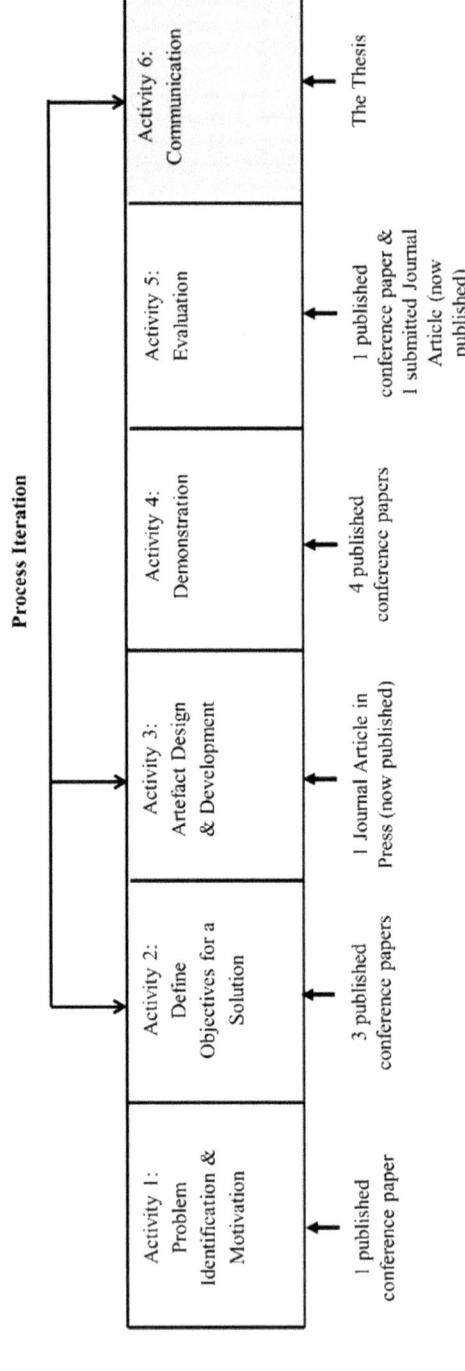

Figure 3.4 Design science research methodology (DSRM) steps to constructing my thesis

about my topic and I started my in-depth literature review. I looked at, and considered, many different paradigms that had previously been used in my topic area, but none were a natural fit for what I wanted to do.

I had identified a business problem that was important enough to warrant a new solution. However, all the approaches in the literature were limited and, in the main, subjective. The DSR paradigm solved everything, I read the 2007 article by Peffers *et al.*, and then the 2004 article by Hevner *et al.* These seminal articles set me on my journey of discovery. There are many articles now available on the DSR paradigm and DSRM, but I recommend you begin with these two.

The DSR paradigm does not require any hypotheses, but you do need well-written research questions. The process is quite logical, and it is iterative, therefore you have the ability to revisit different steps of the process and tweak your artefact or approach if needed. This paradigm really matched my thought processes and approach to research.

As technology is a part of almost everything today, this is a fast way of conducting research and being able to implement any changes into your artefact as necessary. Using hypotheses is still relevant in research, but the process is generally too slow when technology is involved. As the DSR paradigm is widely accepted in IS research, understanding of it by researchers in other disciplines opens up more opportunities in cross-disciplinary research.

I plan to use the DSR paradigm again in research, I have, in fact recommended it to other PhD students (from IT and business) who have also embraced it. I recommend any researcher should consider the DSR paradigm for their research, especially if there is any type of technology (including the internet or digital devices) involved, and especially where the research involves a likely problem-solving solution. I also recommend printing out and placing the six activities of the DSRM in a prominent place for easy reference.

You are going to discover attitudes to publishing your research vary according to the discipline you are in, your supervisors, other researchers and thesis or publications examiners. To some, the most important thing is to publish in top journals. However, this can be a very long process, on average two or three years to publication. A quick top journal publication rarely arises in under 12 months after acceptance of the article.

Although the DSR paradigm has become accepted in IS journals, there may be some resistance from business discipline journals. This resistance may not be directly related to the DSR paradigm itself, but more to do with a lack of knowledge of researchers in this discipline, and in some cases, resistance to change. Therefore, searching out cross-disciplinary journals and those more progressive, and accepting of new ideas, paradigms, and methodologies from other disciplines can be well worth the effort.

In some disciplines (particularly business), publishing in conferences is viewed as having less prestige than publishing in a journal. However, you are getting your research out in the domain early, having it peer reviewed and getting quick constructive feedback right from the very beginning of your research. Using the traditional thesis approach, the only early feedback you are going to receive is from your supervisory team – as you likely may not have much of your research published until well after your thesis has been submitted for evaluation. The risk here is another researcher may have been working in the same area and publishing while you are still writing chapter by chapter. This could mean that by the time you get to write that journal article, your research may be outdated and no longer seen as relevant.

It is important for the PhD candidate and the supervisory team to be early in pursuing and checking the skills sets of potential DSRM thesis examiners. Thesis by publication is often relatively new to many potential global PhD thesis examiners. Hence, to avoid unnecessary criticism and unnecessary comparisons with the traditional thesis approach, I believe for DSRM examinations, only well-experienced (and well-published) examiners in DSR and DSRM should be approached.

I followed the path of conference publications, which proved very useful. The feedback from peer reviewers was extremely beneficial – it improved my writing skills, it enabled me to explain succinctly and clearly my research, provided suggestions of useful articles to read and even future directions my research might take. I have nine conference publications across management, information systems, tourism, e-business, business and decision science. These conferences all accepted my use of the DSR paradigm.

If I had known at the beginning of my PhD journey what I know now I would have looked to other disciplines for a paradigm much earlier than I did. Researchers, both new and seasoned, should be encouraged to look to newer paradigms for their studies. I hope this chapter inspires some of its readers to embrace the DSR paradigm as I have done.

References

Betzing, J.H., Beverungen, D. and Becker, J. (2018) Design principles for co-creating digital customer experience in high street retail. In P. Drews, B. Funk, P. Niemeyer and L. Xie (eds) *The Conference on Information Systems, Data driven X – Turning Data into Value* (pp. 2083–2094). Luneburg, Germany: Leuphana University.

Cassidy, L. and Hamilton, J. (2011a) Website benchmarking: Evaluating scaled and dichotomous approaches. In S. Rotchanakitumnuai and L. Kaewkitipong (eds) The *11th International Conference on Electronic Business, Borderless E-Business for the Next Decade* (pp. 408–412). Bangkok, Thailand: Thammasat University.

Cassidy, L. and Hamilton, J. (2011b) Website benchmarking: theoretical and measurement aspects. In S. Rotchanakitumnuai and L. Kaewkitipong (eds) The *11th International Conference on Electronic Business, Borderless E-Business for the Next Decade* (pp. 413–416). Bangkok, Thailand: Thammasat University.

Cassidy, L. and Hamilton, J. (2012a) Multi-level website benchmarking: Typological collation of recent approaches. In *Proceedings of the 26th Annual Australian and New Zealand Academy of Management Conference* (pp. 1–19). Perth: Australia.

Cassidy, L. and Hamilton, J. (2012b) Website benchmarking: A comprehensive approach. In *Proceedings of the 12th Hawaii International Conference on Business* (pp. 543–548). Honolulu: HICOB, Hawaii.

Cassidy, L.J. and Hamilton, J.R. (2013) A comprehensive approach to capturing website quality measures. In *Proceedings of the 12th International Decision Sciences Institute and 6th Asia-Pacific Decision Sciences Institute Conference* (pp. 280–289). Nusa Dua, Bali: Indonesia.

Cassidy, L.J. and Hamilton, J.R. (2014a) Tropical tourism website qualities. In *Proceedings of the 24th CAUTHE Conference* (pp. 232–245). Brisbane: Australia.

Cassidy, L. and Hamilton, J. (2014b) Location of service industry web objects: Developing a standard. In *Proceedings of the 25th Australasian Conference on Information Systems* (pp. 1–10). Auckland: New Zealand.

Cassidy, L. and Hamilton, J. (2016a) A design science research approach to website benchmarking. *Benchmarking: An International Journal* 23 (5), 1054–1075.

Cassidy, L.J. and Hamilton, J. (2016b) Website benchmarking: An abridged WAM study. *Benchmarking: An International Journal* 23 (7), 2061–2079.

Cassidy, L., Hamilton, J. and Tee, S. (2014) Generating return visitor website traffic. In E.Y. Li (ed.) *The 14th International Conference on Electronic Business and The 1st Global Conference on Internet and Information Systems: Creating Business Values through Innovations in Clouds Services* (pp. 126–133). Taiwan: National Chengchi University.

Cassidy, L., Hamilton, J. and Tee, S. (2015) Generating first time visiting consumer traffic: A live case study. In *Proceedings of the 25th Annual CAUTHE Conference* (pp. 88–100). Gold Coast: Australia.

El-Masri, M. and Tarhini, A. (2015) A design science approach to gamify education: From games to platforms. In *Proceedings of the 23rd European Conference on Information Systems (ECIS 2015) Research-in-Progress Papers* (pp. 1–10). Munster: Germany.

Hevner, A. and Chatterjee, S. (2010) *Design Research in Information Systems: Theory and Practise*. New York: Springer.

Hevner, A.R., March, S.T., Park, J. and Ram, S. (2004) Design science in information systems research. *MIS Quarterly* 28 (1), 75–105.

Hevner, A., vom Brocke, J. and Maedche, A. (2019) Roles of digital innovation in design science research. *Business Information Systems Engineering* 61 (1), 3–8.

Hovorka, D. and Germonprez, M. (2010) Identification-interaction-innovation: A phenomenological basis for an information services view. In D.N. Hart and S.D. Gregor (eds) *Information Systems Foundations: The Role of Design Science* [The Australian National University version] (pp. 3–20). Retrieved from http://epress.anu.edu.au/is_foundations_citation.html

Iivari, J. (2007) A paradigmatic analysis of information systems as a design science. *Scandinavian Journal of Information Systems* 19 (2), 39–64.

Kao, H.-Y., Yu, M.-C., Masud, M., Wu, W.-H., Chen, L.-J. and Wu, Y.-C. J. (2016) Design and evaluation of hospital-based intelligence system (HBIS): A foundation for design science research methodology. *Computers in Human Behavior* 62 (September 2016), 495–505. http://dx.doi.org/10.1016/j.chb.2016.04.021

Kuechler, B. and Vaishnavi, V. (2011) Promoting relevance in IS research: An informing system for design science research. *Informing Science: The International Journal of an Emerging Transdiscipline* 14, 125–138.

Peffers, K., Tuunanen, T., Rothenberger, M.A. and Chatterjee, S. (2007) A design science research methodology for information systems research. *Journal of Management Information Systems* 24 (3), 45–77.

Simon, H.A. (1981) *The Sciences of the Artificial* (2nd edn). Cambridge, MA: MIT Press.

Turetken, O., Grefen, P., Gilsing, R. and Adali, O.E. (2019) Service-dominant business model design for digital innovation in smart mobility. *Business Information Systems Engineering* 61 (1), 9–29.

Vaishnavi, V. and Kuechler, W. (2004) *Design Science Research in Information Systems*. Retrieved from www.desrist.org/design-research-in-information-systems.

4 An Application of Quasi-Experiments to Study Humour in Tourism Settings Guided by Post-Positivism

Anja Pabel

Introduction

The focus of this chapter is on adopting a quasi-experimental research design guided by the post-positivist paradigm to investigate humour in tourism settings. The overall aim of my PhD thesis was directed at better understanding tourists' responses to humour. To achieve this aim, I conducted a quasi-experimental field study using a self-administered questionnaire. The intent of this quasi-experimental field study was to measure the effect that changing humour scenarios had on tourists and to investigate how key variables related to one another. I conducted this study at two tourism settings with different experiences on offer. This also allowed me to undertake a comparative analysis to reveal any differences in tourists' responses to the planned extension of humour. This chapter starts by reviewing the main assumptions of the post-positivist paradigm, addresses in detail why I selected a quasi-experimental research design and reflects on the decisions that needed to be made along the way.

A Review of the post-positivist Paradigm and its Main Assumptions

The post-positivist research paradigm emerged in the 1960s. It acknowledges that there are more ways of knowing aside from using the scientific method proposed by the positivist paradigm (McGregor & Murnane, 2010; Onwuegbuzie & Leech, 2005). The paradigm is called post-positivism because it signifies knowledge creation 'after' positivism and challenges the notion of an absolute truth of knowledge (Phillips & Burbules, 2000). Moreover, it recognises that researchers cannot be 'positive' about their claims of knowledge when studying human behaviours (Creswell, 2008).

In comparing the two paradigms of positivism and post-positivism, the extant literature offers some interesting discussions. For example, some researchers consider post-positivism as a strand of positivism (Ponterotto, 2005) or a milder form of positivism (Willis, 2007), while Zammito (2004) positions post-positivism as a paradigm that denies positivism, i.e. as a non-positivism paradigm. McGregor and Murnane (2010: 421) state that the natural sciences are known for using the positivist paradigm because 'the only way people can be positive that knowledge is true is if it was created using the scientific method.' In contrast, the social and behavioural sciences tend to use the post-positivist paradigm since it recognises the voice of the researcher and the role of the research participants in creating knowledge. This is why the post-positivist paradigm has also been described as a scientific method that has been modified for the social sciences (Taylor & Medina, 2013).

As a philosophy, post-positivism aims to produce generalisable knowledge about social patterns by identifying and assessing the causes that influence outcomes (Creswell, 2003). However, rather than assuming there is a single, tangible reality which is objective and independent of a researcher, post-positivists acknowledge that reality does exist, but it can only be partly known or measured (Ponterotto, 2005). Therefore the absolute truth can never be found and any established research evidence and its inferred causes will always be imperfect and fallible (Phillips & Burbules, 2000).

Its epistemology is mostly based on deductive or reductionist reasoning, in that ideas are reduced into a discrete set of variables to test the relationships among them (Creswell, 2003). Quasi-experiments are a frequently used research method. While quasi-experiments use pre-defined variables, which evaluate the causal impact of a treatment or manipulation on a target population, there is no random assignment of subjects to groups (DiNardo, 2008; Taylor & Medina, 2013). Quasi-experiments are dependent on the context in which the research takes place, and for this reason random assignment of subjects may not always be possible due to ethical and/or practical constraints (Fife-Schaw, 2000; Shadish *et al.*, 2002).

The purpose of post-positivist enquiry is to search for meaning in specific social or cultural contexts by seeking patterns and commonalities or by exploring underlying structures, rather than searching for general laws which may be applicable to everyone (McGregor & Murnane, 2010). Post-positivist researchers recognise that objectivity can never be fully achieved, and knowledge creation cannot be completely free of the researcher's own values, feelings, background knowledge and judgement system, which influences what is studied and how it is studied (Chilisa & Kawulich, 2012). Rather, post-positivism acknowledges that reality is constructed based on the voices of the researcher and the research participants who are involved in creating knowledge. As such, research cannot be value-free, objective and unbiased but is assumed to be value-laden, subjective and intersubjective. Post-positivists recognise that knowledge cannot be isolated from human beings and their lived experiences (McGregor & Murnane, 2010). This is also the reason why quasi-experiments tend to be conducted in the real-life contexts that are important to answering the research questions or in the daily lives of the people being studied, instead of isolating them in laboratory settings. Hence, knowledge creation is dependent on the particular context in which the research takes place.

Data gathering instruments can be based on quantitative and qualitative methodologies and may include quasi-experiments, survey research, mind maps, public consultations, interviews, participant-observation and causal comparative research designs (Creswell, 2008; McGregor & Murnane, 2010). As mentioned before, the focus is on quasi-experimental designs since post-positivist researchers are aware that purely experimental designs are difficult to achieve in social science contexts like tourism or consumer behaviour due to the subjective realities of individuals (Chilisa & Kawulich, 2012; McGregor & Murnane, 2010).

Validity in quantitative studies is an important consideration when designing a research project since it shows that the findings truly represent the phenomenon that is measured. Quasi-experiments offer high levels of external validity because they are conducted in the real-life contexts that matter to the research project (Viglia & Dolnicar, 2020). However, internal validity may be a concern due to two reasons. Firstly, any control or comparison groups may not be similar at baseline (Rossi *et al.*, 2004). Secondly, the process of assigning treatment conditions depends on the researcher who may be influenced by their own biases (DiNardo, 2008). Hence the degree to which the results are attributable to a treatment or manipulation are likely to be imperfect and fallible, as is acknowledged by the post-positivist worldview. Furthermore, large sample sizes are required to create generalisable knowledge about social patterns (Batistatou *et al.*, 2014; Creswell, 2012). Nevertheless, Baldwin and Berkeljon (2010) assert that careful planning of quasi-experiments allows for strong causal inferences.

Application of the Paradigm in my PhD Research

In this section, I'm describing the logistics of running quasi-experiments that were part of my PhD research. From my readings, I knew that tourism businesses applied humour differently depending on the various contexts in which they operate (Frew, 2006; Pearce, 2008, 2009). For example, humour may be included to make tourism audiences feel more comfortable in a new setting, to entertain them, to make factual information more relatable or a combination of these reasons. What I did not know was whether any of these aspects could be enhanced in response to a planned extension of humour. I also must admit that I had always wanted to conduct a 'proper' experiment to uncover whether, and to what extent, a certain intervention or manipulation influences one or more outcome variables.

When I started my PhD research into the humour–tourism relationship, I thought it would be a great idea to design an experiment using functional magnetic resonance imaging (fMRI) to measure responses to humour stimuli, similar to the previous research that I had read about (Bartolo *et al.*, 2006; Sawahata *et al.*, 2013). Eventually, I realised that such an approach was not very realistic to the context in which most tourism experiences take place and there were several challenges. One issue was the cost of accessing an fMRI and then there were also challenges with accessibility; that is, recruiting and inviting tourists to come to an fMRI facility to measure their responses to different humour scenarios. On further consideration, I also realised that such an approach would have taken me away from a post-positivist direction since purely experimental designs are predominantly guided by a positivist worldview. Furthermore, post-positivist researchers are aware that true experiments are difficult to achieve in social science contexts due to the subjective realities of individuals (Chilisa & Kawulich, 2012; McGregor & Murnane, 2010).

Since I was still keen on conducting an experiment and was aware that I needed to think of a more naturalistic way of quantitatively measuring the humour responses of tourists while they were visiting an actual tourism attraction, I went back to the literature. During this time, I started to read more about naturalistic ways of conducting studies and I came across quasi-experiments. It was then that I had 'my light-bulb moment' and made the decision to conduct a field-based quasi-experiment to measure the effect that changing humour treatment scenarios had on tourists.

Description of my method

I conducted a cross-sectional, field-based, quasi-experimental study at two tourism settings employing a survey-based questionnaire to measure tourists' responses. Cross-sectional designs provide information about respondents' opinions and attitudes at one point in time (Creswell, 2012). The cross-sectional design made it possible to compare responses between groups and between tourism settings. At this point it might also be helpful to reiterate the differences between true experimental designs and quasi-experimental designs. True experiments are characterised by random assignment of respondents to different groups or treatments, which is not a requirement for quasi-experiments (Creswell, 2012; Goldstein & Renault, 2004).

When conducting true experiments, randomisation based on statistical probability ensures that 'there should be no systematic difference on any rival causal factors between the cases receiving the treatment and those that do not' (Goldstein & Renault, 2004: 738). However, there are times when it is impractical or unethical to randomly assign participants to treatment conditions (Creswell, 2012; Fife-Schaw, 2000; Goldstein & Renault, 2004). There could also be practical considerations which can limit the researcher's control over some situations (Fife-Schaw, 2000). For example, in most tourism settings it is up to tourists themselves, and not a researcher, to decide what activities or shows they would like to attend during their tourism experience. It would clearly be inappropriate to tell them they could not be part of certain activities or shows because of research-based needs about sampling and statistical probability. Post-positivist enquiry recognises the context in which the research takes place, and therefore random assignment of subjects to quasi-experiments is not always possible due to practical constraints (Fife-Schaw, 2000; Shadish et al., 2002).

Furthermore, it is common for quasi-experiments to be conducted in the field as opposed to a laboratory setting. However, researchers operating in real-life settings often have limited control over segregating or minimising confounding variables which might influence the relationship between the other variables (Creswell, 2012; Fife-Schaw, 2000). In regard to confounding variables which might have influenced this study, it needs to be recognised that as the researcher, I had no control over aspects such as the respondents' individual humour appreciation, their attitudes towards humour and prior experience with other tourism experiences. Any of these aspects could have influenced the participants' responses to the questionnaire in this study.

Naturalistic quasi-experimenting

Lee (2000: 142) defines a field experiment as 'an experiment conducted outside the laboratory context.' Field experiments performed in their natural setting give the researcher better opportunities to observe naturally occurring events and behaviours. It is still possible to introduce an intervention or treatment into a field-based location to observe and measure its consequences and to reveal any interesting relationships (Tunnell, 1977; Weick, 1968). However, it is imperative that large-scale manipulations, that could cause too much disturbance, are avoided to keep a study as natural as possible. Weick (1968) referred to this as 'tempered naturalness' where a natural setting is only slightly modified to observe how behaviours are changing.

Tunnell (1977) mentioned three dimensions to keep field research as natural as possible, including natural setting, natural behaviour and natural treatment, each of which bring a research project closer to the real world. *Natural setting* refers to any setting outside the laboratory where people would naturally gather. *Natural behaviour* is characterised through naturally occurring behaviour and responses, such as behaviour that would exist even without the experimental manipulation taking place. *Natural treatment* refers to naturally occurring discrete events to which the research participants are exposed. The more of these natural dimensions are combined in a given research study, the greater the degree of generalisability of the research findings (Tunnell, 1977).

These natural research dimensions were also applied to my PhD research. The natural setting was guaranteed through the background situation of the chosen tourism settings where the study took place. The natural treatment was represented by the manipulations utilised in this study where participants were exposed to changing humour treatment scenarios. The natural behaviour was measured by participants' reactions to humour on various scales.

Adhering to the three natural dimensions of research proposed by Tunnell (1977) has several advantages. First, the findings of the research are bound to be more meaningful because they were conducted in their real-world setting. Second, combining all three natural dimensions adds not only richness to the research but also improves its external validity (Tunnell, 1977). While it is not advised to generalise findings from any one study, the use of real-life settings, behaviours and treatments make the research findings more applicable to other settings than research conducted in laboratories (Flyvbjerg, 2001; Tunnell, 1977). Of course, the specific and individual contexts of some settings make full generalisability a misplaced undertaking, especially considering that social, cultural and historical changes continue to take place in real-life settings.

Description of tourism settings and questionnaire design

The naturalistic cases in this study were two real-life, commercial tourism settings especially chosen because they were already successful in including humour in their tourism presentations. This was ascertained by reading through comments on the travel review website Tripadvisor, which were used as public acknowledgement that the businesses were successful in their application of humour.

One attraction is based on zip-lining in a nature-based setting. The experience is offered eight times daily and lasts approximately two hours. Every tour is limited to 13 people. The company's brochure promises that the experience will be 'fun and educational.' The second tourism operator is a wildlife park focusing on crocodile conservation. The park offers educational and entertaining presentations including crocodiles, snakes, cassowaries, koalas and other wildlife at several times throughout the day.

I used a questionnaire in this study to measure the outcome of the *natural treatment*. Questionnaires were most appropriate in this context because they could be distributed to larger samples at the chosen tourism settings. The questionnaire was designed to be as short as possible considering that this study took place at real-world tourism attractions. It was handed out at the end of respondents' tourism experiences as they were exiting or preparing to leave the setting. Questions measured how respondents perceived the humour they experienced in the context of the two settings. For example, a total of eight scales were used to collect information about the respondents' awareness of the humour as well as their satisfaction rating.

Initially the questionnaire was pilot tested to ensure that research participants would have no difficulty in understanding the questions and were clear on how to complete the instrument (Creswell, 2012). I handed the questionnaire to five friends with special instructions to note down any ambiguities and mistakes. Four friends had previously been to either one or even both of the tourism settings chosen in this study and therefore had experience with these tourism settings. Based on the comments received, I made some appropriate but quite limited changes.

Manipulation of the treatment scenarios

During my quasi-experimental field research, I exposed tourists to changing humour scenarios to document their effects. The treatment variable were different humour scenarios, which attempted to measure any effects that potential increases in humour would have on the dependent outcome variables of humour responses. The dependent variables were measured on scales; for example, comfort, concentration, connection, satisfaction.

When working with different manipulation scenarios, it is important to have an appropriate control or comparison group to ensure that unusual effects are not due to peripheral factors (Mark, 2010). The quasi-experimental design in this study consisted of several phases. The first phase was regarded as a comparison condition since no experimental manipulation was applied. This allowed me to see how respondents would typically respond to the humour provided by the tour guides. Having such a baseline situation also acknowledged that humour was already a natural occurrence at the two chosen tourism settings.

In an attempt to experimentally increase the humour, the second phase included asking tour guides to be as funny as they possibly could through their own initiated humorous efforts. The third phase involved increasing humour efforts by asking tour guides to add humorous material sourced by me into their tourism scripts. Tour guides were prompted to remember and use five of the comments with which they were most comfortable. I expected that such a quasi-experimental set-up would allow me to make comparisons between the three treatment scenarios.

The humorous material given to the guides was sourced from the internet from joke collection websites and websites outlining funny rainforest related facts. This proved to be a more difficult undertaking than I anticipated. Finding humorous material which could be handed to tour guides was a lengthy process and did not include many direct hits. Keeping this in mind, it can be recognised how challenging it is for tour guides to source their own humorous materials. Once a good selection of humorous comments and facts were found, the material was checked for its appropriateness by a panel of judges including two tour guides, two academics and two lay people who stated there was no offensive materials included. The materials were judged to be reasonably good examples of jokes, puns and nonsense humour.

On-site procedures

The questionnaire was administered immediately following the tourism experiences ensuring that tourists' impressions of the humour they experienced during on-site presentations were still fresh. Convenience sampling was used for the actual distribution of questionnaires on-site. Creswell (2012) defines convenience sampling as a technique where research participants are selected because they are willing and available to take part in a study.

Since this sampling technique is non-random, it is impossible to say how representative a sample is of the population, but the responses should still provide useful information in answering the set research questions (Creswell, 2012). Employing convenience sampling enabled me to make contact with a large number of respondents quickly and cost effectively (Hair *et al.*, 2003). In terms of sample size itself, Creswell (2012) recommended to collect data from as large a sample as possible because then there would be less chance that the selected sample is different from the population. Considering this suggestion, the data were collected during school holidays to maximise response opportunities. The number of questionnaires collected in this study was linked to the number of scenarios used. I considered 100 completed questionnaires per scenario was sufficient to run appropriate statistical analysis. This concurs with the literature in that Hair *et al.* (2014: 10) state that 'a moderate sized power reaches acceptable levels at sample sizes of 100 or more for alpha levels of both 0.05 and 0.01'.

Table 4.1 provides details on the number of the manipulation scenarios and outlines the number of responses collected for each scenario. Data collection at the wildlife park included only two manipulation scenarios because at this study location it was my intention to produce a stronger manipulation by combining manipulation scenarios 1 and 2 used at the zip-lining attraction into one treatment scenario. However, the head tour guide at the wildlife park was not keen to use the humorous material and decided that they would merely try to be as funny as they could be.

Analysis of questionnaire data

SPSS (version 20) and Leximancer (version 4) were used to analyse the collected data. Descriptive statistics were used to highlight any overall tendencies in the data. To make comparisons between groups, settings

Table 4.1 Data collection at the two tourism attractions

	Baseline scenario	Manipulation scenario 1	Manipulation scenario 2
Zip-lining attraction	Measure responses to humour as it typically occurs during the tour	Tour guides asked by researcher to be as funny as they can be	Tour guides asked to use additional humorous material sourced by researcher
Number of responses	103	107	100
Wildlife park	Measure responses to humour as it typically occurs during the tour	Tour guides asked by researcher to be as funny as they can be	
Number of responses	101	103	

and the various manipulation scenarios, inferential statistics were used to examine the influence of the independent variable on the dependent variables. Correlational statistical tests were used to describe the degree to which variables relate to one another. Presenting the results of this study is beyond the scope of this chapter. Readers who are interested in the results are directed to Pabel and Pearce (2016).

Reflections on my Experience in Working within a Post-positivist Worldview

In reflecting on my experience with the post-positivist paradigm and using the quasi-experimental design, I have learned that research does not always go according to plan, irrespective of how good a plan one has. For example, when designing the study, it was 'my anticipation' that humour perception would indeed increase significantly with the changing humour scenarios. However, none of the changes created through the manipulation scenarios showed a statistically significant result at the zip-lining attraction. In fact, the humour scale item showed a downward trend between the different manipulation scenarios while the satisfaction scale item stayed unchanged.

After realising that the manipulation scenarios did not work according to my anticipation, I decided after discussions with the supervisory team, that it would be worthwhile to see if a similar outcome would be the result at a second tourism operation. This is why the study was then also undertaken at the wildlife park. This time, however, the manipulation was simplified and included only two scenarios instead of three, due to reservations of the head tour guide, which are explained below. The first scenario included tourists being exposed to the normal scripts that tour guides usually used during their presentations. The second scenario included asking the tour guides to be as humorous as they can be during their speeches to the tourists. The trend at the wildlife park for both the perceived level of humour scale item and the satisfaction scale item displayed an upward tendency indicating that humour was potentially successful in creating more enjoyable and satisfying on-site tourism experiences. Yet again, none of the results were statistically significant.

As a 'researcher in training', I could not help but wonder 'Where did I go wrong?' Were the humorous materials which I provided to the tour guides not funny enough? Were there too many confounding variables at play that influenced humour perceptions, e.g. weather on the day, individual differences in humour appreciation, respondents' attitudes towards humour and prior experience with other tourism experiences. In trying to explain why the manipulation scenarios did not have the hypothesised enhancing effect, I reflected on the following points.

Firstly, it needs to be acknowledged that individual differences exist in regard to people's humour appreciation based on their gender, age, nationality, previous experiences and many other factors. Secondly, while the tour guides showed goodwill in using the provided humorous material in their tourism presentations, there may have been situations that arose during the tours that made it impossible to be consistent in their humour effort across different contexts. A further explanation would be that the instructions given to the tour guides may have been ambiguous or not clear enough. From the onset of this study I recognised that I only had limited control over how humour could be manipulated and perceived. My study relied on the tour guides' goodwill, and it is challenging to say to what extent my instructions were followed by the tour guides. Therefore, good liaisons with the management staff are necessary when organising similar quasi-experimental studies in the future that highlight exactly what is required.

Conducting research in real commercial settings can be problematic at times, as shown at the wildlife park, where the head tour guide had reservations about including the humorous material sourced by me. His reasoning was that humour was merely used to add to the experience and that their focus was never on delivering outright jokes because if jokes fell flat then it would be difficult to recover and reconnect with the audience. Hence, incorporating various measures may be useful for the research project but might be less suitable for the real-life tourism settings where the studies are conducted. In this case it led to restrictions in how the study was carried out due to the head tour guide's concern. Practical issues when working in real commercial tourism environments have also been acknowledged by Reiser and Simmons (2005) who note that quasi-experimental studies can be complex, include design constraints, lack of control over various external background factors and are often a time and resource consuming process. However, I still acknowledge the importance of using a quasi-experimental study design irrespective of certain constraints because study results are closer to real-world contexts rather than results established in laboratory settings.

There are several considerations to be pondered prior to conducting a quasi-experiment including context-specific restrictions and user participation (Adèr et al., 2008; Parry et al., 2001). For example, in the context of my research, I would have liked to design a survey that involved pre- and post-intervention evaluation by the same research participants, however such research designs are not practical in real-world tourism settings. When conducting quasi-experiments guided by the post-positivism paradigm, context is everything and so I was mindful that having to conduct pre- and post-intervention surveys could contribute to tourists becoming annoyed at the tourist attraction. Clearly, tourists enter an attraction to have a good time or to spend quality time with their family and friends, and completing a survey is not a top priority for them.

Conclusion

Evidence-based research in tourism is designed to provide new knowledge and to enhance the outcomes for tourists or tourism workers.

What is important to consider though is that sound research should fit the context of the real-world tourism settings in which it is conducted. In undertaking meaningful research, many decisions need to be made regarding which research design best answers the research question. Post-positivism as a guiding research paradigm, grew from the positivist paradigm with its emphasis on scientific methods, but provides researchers with more flexibility because it focuses on contexts, culture and subjectivity (Creswell, 2008). In this chapter, I reflected on my experience with post-positivism as well as my anticipations and restrictions in conducting a quasi-experimental field study. Quasi-experimental research designs can lead to interesting discoveries. In the context of my PhD research I must confess that I enjoyed the process of setting up a quasi-experiment. It was challenging, yet fun, to develop humorous materials that tour guides might include in their presentations to tourists. I hope that outlining my research processes will assist emerging scholars in designing their own quasi-experiments and to develop original and interesting projects.

References

Adèr, H.J., Mellenbergh, G.J. and Hand, D.J. (2008) *Advising on Research Methods: A Consultant's Companion*. Huizen: Johannes van Kessel Publishing.

Baldwin, S. and Berkeljon, A. (2010) Quasi-experimental design. In N.J. Salkind (ed.) *Encyclopedia of Research Design* (pp. 1171–1175). Thousand Oaks, CA: SAGE Publications.

Bartolo, A., Benuzzi, F., Nocetti, L., Baraldi, P. and Nichelli, P. (2006) Humour comprehension and appreciation: An fMRI study. *Journal of Cognitive Neuroscience* 18 (11), 1789–1798.

Batistatou, E., Roberts, C. and Roberts, S. (2014) Sample size and power calculations for trials and quasi-experimental studies with clustering. *The Stata Journal* 14 (1), 159–175.

Cook, T.D. and Campbell, D.T. (1979) *Quasi-experimentation: Design and Analysis Issues for Field Settings*. Boston, MA: Houghton Mifflin.

Chilisa, B. and Kawulich, B. (2012) Selecting a research approach: Paradigm, methodology and methods. In C. Wagner, B. Kawulich and M. Garner (eds) *Doing Social Research: A Global Context* (pp. 51–61). New York: McGraw-Hill Higher Education.

Creswell, J.W. (2003) *Research Design: Qualitative, Quantitative, and Mixed Methods Approaches* (2nd edn). Thousand Oaks: SAGE Publications.

Creswell, J. (2008) *Educational Research: Planning, Conducting and Evaluating Quantitative and Qualitative Research* (3rd edn). Upper Saddle River, NJ: Pearson Prentice Hall.

Creswell, J.W. (2012) *Educational Research: Planning, Conducting and Evaluating Quantitative and Qualitative Research* (4th edn). Boston, MA: Pearson Education.

DiNardo, J. (2008) Natural experiments and quasi-natural experiments. In S.N. Durlauf and L.E. Blume (eds) *The New Palgrave Dictionary of Economics* (2nd edn). Basingstoke: Palgrave Macmillan.

Fife-Schaw, C. (2000) Quasi-experimental design. In G.M. Breakwell, S. Hammond and C. Fife-Schaw (eds) *Research Methods in Psychology* (pp. 75–87). London: Sage Publications.

Flyvbjerg, B. (2001) *Making Social Science Matter: Why Social Inquiry Fails and How it Can Succeed Again*. Cambridge: Cambridge University Press.

Frew, E. (2006) Humorous sites: An exploration of tourism at comedic TV and film locations. *Tourism Culture and Communication* 6, 205–208.
Goldstein, H. and Renault, C. (2004) Contributions of universities to regional economic development: A quasi-experimental approach. *Regional Studies* 38 (7), 733–746.
Hair, J.F., Babin, B., Money, A.H. and Samouel, P. (2003) *Essentials of Business Research Methods*. Hoboken, NJ: Leyh Publishing, LLC.
Hair, J.F., Black, W.C., Babin, B.J. and Anderson, R.E. (2014) *Multivariate Data Analysis* (7th edn). Harlow: Pearson Education Limited.
Lee, R.M. (2000) *Unobtrusive Methods in Social Research*. Buckingham: Open University Press.
Mark, M.M. (2010) Emergence in and from quasi-experimental design and analysis. In S.N. Hesse-Biber and P. Leavy (eds) *Handbook of Emergent Methods* (pp. 87–108). New York: The Guilford Press.
McGregor, S.L.T. and Murnane, J.A. (2010) Paradigm, methodology and method: Intellectual integrity in consumer scholarship. *International Journal of Consumer Studies* 34, 419–427.
Onwuegbuzie, A.J. and Leech, N.L. (2005) Taking the "Q" out of research: Teaching research methodology courses without the divide between quantitative and qualitative paradigms. *Quality and Quantity* 39, 267–296
Pabel, A. and Pearce, P.L. (2016) Tourists' responses to humour. *Annals of Tourism Research* 57, 190–205, doi: 10.1016/j.annals.2015.12.018.
Parry, O., Gnich, W. and Platt, S. (2001) Principles in practice: reflections on a 'postpositivist' approach to evaluation research. *Health Education Research Theory Practice* 16 (2), 215–226.
Pearce, P.L. (2008) Studying tourism entertainment through micro-cases. *Tourism Recreation Research* 33 (2), 151–163.
Pearce, P.L. (2009) Now that is funny: Humour in tourism settings. *Annals of Tourism Research* 36 (4), 627–644.
Phillips, D.C. and Burbules, N.C. (2000) *Postpositivism and Educational Research*. Lanham, MD: Rowman & Littlefield.
Ponterotto, J.G. (2005) Qualitative research in counselling psychology. *Journal of Counseling Psychology* 55, 126–136.
Reiser, A. and Simmons, D.G. (2005) A quasi-experimental method for testing the effectiveness of ecolabel promotion. *Journal of Sustainable Tourism* 13 (6), 590–616.
Rossi, P.H., Lipsey, M.W. and Freeman, H.E. (2004) *Evaluation: A systematic approach* (7th edn). London: SAGE.
Sawahata, Y., Komine, K., Morita, T. and Hiruma, N. (2013) Decoding humour experiences from brain activity of people viewing comedy movies. *PLoS ONE* 8 (12), e81009.
Shadish, W.R., Cook, T.D. and Campbell, D.T. (2002) *Experimental and Quasi-experimental Designs for Generalized Causal Inference*. Boston: Houghton Mifflin Company.
Taylor, P.C. and Medina, M. (2013) Educational research paradigms: From positivism to multiparadigmatic. *Journal for Meaning-Centered Education* 1, 1–16.
Tunnell, G.B. (1977) Three dimensions of naturalness: An expanded definition of field research. *Psychological Bulletin* 84 (3), 426–437.
Viglia, G. and Dolnicar, S. (2020) A review of experiments in tourism and hospitality. *Annals of Tourism Research* 80. DOI: https://doi.org/10.1016/j.annals.2020.102858.
Weick, K.E. (1968) Systematic observational methods. In G. Lindzey and E. Aronson (eds) *The Handbook of Social Psychology* (2nd edn, pp. 357–451). Reading, MA: Addison-Wesley.
Willis, J.W. (2007) *Foundations of Qualitative Research: Interpretive and Critical Approaches*. Thousand Oaks, CA: Sage Publications.
Zammito, J.H. (2004) *A Nice Derangement of Epistemes: Post-positivism in the Study of Science from Quine to Latour*. Chicago: University of Chicago Press.

5 Knowledge Co-Production in Tourism and the Process of Knowledge Development: Participatory Action Research

Tramy Ngo, Gui Lohmann and Rob Hales

Introduction

Community-based tourism enterprises (CBTEs) are micro-scaled, rural and remote area-located businesses where the local community is the owner, the manager and the primary beneficiary. Although CBTEs have proliferated in less developed countries because of their potential to transform sustainable tourism objectives into actions, only a few CBTEs can claim to be successful (Goodwin & Santilli, 2009; Rocharungsat, 2008). Most CBTE projects experience market failure and collapse after the initial funding period (Dixey, 2008; Mielke, 2012). In studies of CBTEs, a collaborative approach is widely advocated to help CBTEs overcome their marketing challenges and shift their businesses towards long-term success (Dodds *et al.*, 2016; Idziak *et al.*, 2015; Mbaiwa *et al.*, 2011). Accordingly, there is a need to develop a collaborative marketing approach to underpin stakeholder relationships in marketing co-efforts for the business sustainability of CBTEs. The development of such an approach requires the involvement of multiple stakeholders with different perspectives. To this end, collaborative forms of knowledge generation are proposed (Carr *et al.*, 2016; Torres-Delgado & Saarimen, 2014). In particular, where subaltern viewpoints and marginalised voices are involved, these should be integrated into knowledge generation (Chambers & Buzinde, 2015). Owing to the recognition of different forms of knowledge (Berkes, 2009), a knowledge co-production approach is appropriate to frame the development of a collaborative marketing approach for CBTE sustainability. The knowledge co-production approach is governed by the constructivist paradigm and is facilitated in research processes through participatory action research – PAR (Castleden *et al.*, 2012; Espeso-Molinero *et al.*, 2016).

Constructivism as a Research Paradigm

A research paradigm consists of 'the basic belief system or worldview that guides the investigator, not only in the choice of the method but in ontological and epistemological fundamental ways' (Guba & Lincoln, 1994: 105). A research paradigm reflects the standpoints of researchers regarding ontological, epistemological and methodological stances. Accordingly, a research paradigm underpins the involvement of researchers and research participants in the research process, the tools and techniques of data collection and analysis, the interpretation of research outcomes and the evaluation of research contributions. According to Guba and Lincoln (1994), the selection of a specific research paradigm is dependent on the answers to ontological, epistemological and methodological questions, which include, 'What is the form and nature of reality and, therefore, what is there that can be known about it?'; 'What is the nature of the relationship between the knower or would-be-knower and what can be known?'; and 'How can the inquirer (would-be-knower) go about finding whatever he or she believes can be known?'. Answering the ontological question determines the answers to the epistemological and methodological questions and shapes a selected research paradigm.

As the domain of collaborative marketing for the business sustainability of CBTEs involves CBTE stakeholders of diverse viewpoints and the study on this topic comprises researchers and practitioners from different ways of knowing, a constructivist approach is adopted as the research paradigm. The constructivist approach, according to Guba (1990), admits that multiple realities exist in the minds of the 'insiders' and attempts to obtain one or a few constructions that are reconciled from different perspectives. Multiple realities co-exist because of the interpretation of viewpoints belonging to different cosmologies. A constructivism-based study should focus on the values constructed at the *inter-sphere* of different perspectives. It is argued that this paradigm effectively supports the investigations of highly contextualised problem domains, in which different worldviews co-exist and might be incongruent with each other (Hollinshead, 2006). The constructivist approach enables a harmonising platform for the interactions of diverse viewpoints of CBTE stakeholders and co-researchers for the research goal of co-proposing a sustainable marketing approach for the long-term success of CBTEs. Based on the constructivist paradigm, a co-construction approach to knowledge generation was adopted to guide the research process. The knowledge co-production approach and its applications in the research stages are illustrated in the rest of this chapter.

Knowledge Co-production in Tourism Research

Knowledge co-production refers to an interactive approach to knowledge generation whereby researchers and research participants interact and influence each other to different degrees in generating knowledge, and so the nature of the knowledge generated is considered to be socially constructed knowledge (Dale & Armitage, 2011). Knowledge co-production is based on the premise that knowledge generation is an outcome of collaborative interactions among researchers and research participants. Researchers possess evidence-based perspectives in the academic world, whereas research participants represent diverse viewpoints of the social world. Knowledge co-production allows a research participant to reconcile his or her perspectives with those of others. New knowledge is developed through interactions between researchers and research participants. The new knowledge is characterised by its dual attributes of scientific validity and social relevance (Nowotny, 2003). Moreover, in the realm of knowledge co-production, each research participant is both a knowledge generator and a social learner (Berkes, 2009; Dale & Armitage, 2011; Pohl *et al.*, 2010). Here, participants co-generate knowledge through interactions and concomitantly experience a social learning process through their interactions.

Castleden *et al.* (2012) argue that five attributes define a knowledge co-production study. First, researchers and research participants both share ownership in establishing the study's direction through the various stages of the research progress. The researchers should be flexible in experimenting with different research approaches throughout the research process. Second, the perspectives of both researchers and research participants are regarded as legitimate sources of knowledge. Data collected from the research participants are considered to be an independent school of thought without being subjugated to scientific knowledge. Third, reflexive learning among researchers and research participants is promoted. Fourth, new knowledge is co-generated. Fifth, new knowledge is disseminated in harmony with the culture, values and beliefs of the viewpoints involved to obtain mutual benefits. The practice of these five attributes in a study enables the abovementioned potential of the knowledge co-production approach to be achieved.

The application of a knowledge co-production approach in tourism studies is increasing. Most of the tourism studies using the knowledge co-production approach focus on sustainability; including natural resource management, climate change adaptation and destination sustainability. For instance, numerous studies investigate the significance of the knowledge co-production approach in the management and governance of natural resources in tourist destinations (see Marshall *et al.*, 2016; Ungar & Strand, 2012). Climate change adaptation for destination sustainability is another domain in which the knowledge co-production approach has been employed (see Armitage *et al.*, 2011). The emergence of the knowledge co-production approach within the broad discipline of sustainability can inform tourism studies. Nevertheless, few tourism studies have used the knowledge co-production approach (Becken *et al.*, 2015; Espeso-Molinero *et al.*, 2016). Moreover, the scarcity of tourism research using the knowledge co-production approach contrasts with the recurrent call for its application in tourism studies (Carr *et al.*, 2016; Chambers & Buzinde, 2015).

The majority of knowledge co-production studies focus on the significance of this approach in terms of integrating two different ways of knowing, i.e. indigenous knowledge and scientific knowledge, for the benefit of collaborative knowledge. Current knowledge co-production studies investigate collaborative works between Western-based knowledge and indigenous knowledge (Armitage *et al.*, 2011; Davidson-Hunt & O'Flaherty, 2007; Weiss *et al.*, 2013) and between intellectual works in the North and those in the South (Castleden *et al.*, 2012; Chambers & Buzinde, 2015). However, there is an absence of studies illustrating the potential of a knowledge co-production approach for harmonising practitioner standpoints and researcher viewpoints. This paucity is in contrast to the recent burgeoning of tourism studies arguing for the capacity of the knowledge co-production approach in bridging the research-practice gap (Dredge & Jamal, 2015; Font & McCabe, 2017). Additionally, owing to its acknowledgement of different standpoints and appreciation of the voices of the marginalised, the knowledge co-production approach has the potential to reconcile diverse perspectives in collective efforts. However, it remains an under-utilised approach to the tourism agenda. Thus, by conducting a case study in the domain of CBTE collaborative marketing for business sustainability, we argue that the knowledge co-production approach can responsively address diverse perspectives in collaborative works and shorten the research-practice gap for the benefit of all stakeholders.

At a different level, although many studies investigate the significance of knowledge co-generation in terms of problem-solving (Marshall *et al.*, 2016; Ungar & Strand, 2012) and the outcomes of the knowledge co-production process (Becken *et al.*, 2015; Espeso-Molinero *et al.*, 2016; Holmes *et al.*, 2016), less attention has been paid to illustrating how to develop knowledge through collaborations among knowledge holders. The knowledge development process should encompass the generation of socially constructed, value-based knowledge through interactions among research stakeholders while including those stakeholders' reflections on their learning experience throughout the process. We attempted to address this gap by interrogating the process of knowledge development in a knowledge co-production study focused on CBTE collaborative marketing.

Accordingly, the PAR approach was employed in the investigation process to achieve the aim of this study. PAR is a useful tool for learning promotion and knowledge co-generation (Kindon *et al.*, 2007; Reason & Bradbury, 2008), and therefore, is an essential part of the knowledge co-production process. More specifically, PAR provides a methodological framework for exploring diverse perspectives on CBTE collaborative marketing, bringing them together for knowledge interactions and evaluating learning outcomes achieved through an action learning cycle. Hence, we were confident that the utilisation of PAR complemented and extended our ontological constructivist stance.

PAR in the Knowledge Co-Production Study of CBTE Collaborative Marketing

Participatory action research

PAR is an integrated paradigm of participatory approaches and action-oriented research. More specifically, PAR refers to an inquiry-based approach in which relevant participants in a research project change and improve a problem by actively examining it together (Kindon et al., 2007). Through this research process, researchers and research participants jointly produce knowledge that is understandable, actionable and accessible to them. Thus, participant collaboration and knowledge co-production shape the PAR methodology.

In PAR, the research process is designed through recurring stages of research, action and reflection, as illustrated in Table 5.1 (Kindon et al., 2007). PAR values two types of research outcomes. First, the collective actions and quality information generated are counted as research contributions. Second, the self-mobilisation of research participants in terms of skills, knowledge and capacities throughout the research experience is also evaluated (Kindon et al., 2007). Following this process, collaborative methods and research implementation techniques are employed. Aligning with the flexibility and non-coercive nature of PAR, methodological techniques are very diverse. These techniques can be based on traditional tools such as interviews and focus groups, and on innovative tools using technology; for instance, diagrams and videos (Kindon et al., 2007). Method selection is context-based and considers research participants' capabilities

Table 5.1 Stages of the PAR process

Stages	Activities
Research	The research participants identify a context-pertinent problem that needs to change.
Action	The research participants interact and arrive at a set of steps for change or improvement. The set of actions is context-plausible, and it aligns with the capabilities of research participants.
Reflection	The research participants experience learning and reflection during the action implementation.

Source: Adopted from Kindon et al., 2007.

and resources. Flexible research contributions in PAR and different methodological tools reinforce the prevalence of PAR in the tourism literature.

PAR as a methodological framework

In the realm of tourism, PAR has been used extensively in the context of sustainable tourism research (Cole, 2006; Idziak *et al.*, 2015). PAR is regarded as an effective approach to integrating different viewpoints and multiple disciplines in sustainable tourism, fostering the potential to achieve sustainable tourism in practice and at the grassroots level. The method has been utilised to empower the indigenous communities, which are conventionally viewed as marginalised in the tourism planning process (see Cole, 2006; Idziak *et al.*, 2015). It is also recognised that PAR facilitates individual and social changes towards sustainability orientations (see Jamal & Watt, 2011). Owing to its potential to empowering marginalised voices in the research process and facilitating learning outcomes for participants, PAR is advocated as a base framework to assess the knowledge co-production process.

Indeed, the process of knowledge development through harmonising diverse perspectives in developing a responsive collaborative marketing approach for CBTE's business sustainability can be facilitated by PAR. First, the acknowledgement of diverse viewpoints in PAR allows different types and sources of knowledge to be voiced in the process of knowledge co-production. PAR arguably amplifies the voices of research participants, specifically those of indigenous knowledge holders. Accordingly, indigenous knowledge is recognised and actively engaged in co-generating new knowledge. Second, it is argued that PAR's action-based principle facilitates knowledge interactions, which remain central to the knowledge co-production process. The interactions of different types and sources of knowledge are used to achieve a new socially constructed, value-based knowledge that embraces the viewpoints of different knowledge holders. Third, the PAR approach of learning through engagement paves the way to assessing learning experiences, as reflected by both the researchers and the research participants in the knowledge co-production process. Thus, PAR is used to construct a research design, exploring the process of knowledge development through the collaborative works of researchers and CBTE stakeholders in different stages of the study.

Application of the Paradigm: The Research Context

The context of this study is CBTEs in Vietnam. Specifically, based on the purposive sampling method, three CBTEs in Vietnam were approached: Triem Tay Floating Restaurant, Thanh Toan Gardening and Cookery, and Minh Tho Homestay. These CBTEs represent different development models and diverse marketing approaches for CBTEs in Vietnam. Table 5.2 provides background information on the investigated CBTEs.

Table 5.2 Background information of the investigated CBTEs

Case study	Location	Characteristics
Triem Tay Floating Restaurant	o Triem Tay Village, Quang Nam Province o 3 km from Hoi An City, a tourist centre o The village is facing out-migration issues due to land erosion (ILO, 2015)	o Was launched in June 2015 o Is owned by a Kinh[1] family o Received support from the International Labour Organisation (ILO) and UNESCO for technical training, field trips, marketing and promotion, etc. o Offers food and beverage packages and boating experiences o Is a member of Triem Tay's CBTE cooperative[2], which was established in September 2015
Thanh Toan Gardening and Cookery	o Thanh Toan Village, Thua Thien Hue Province o 8 km from Hue City, a tourist centre o The village is renowned for Thanh Toan Tile-Roofed Bridge, a National Heritage Site and a tourist attraction	o Was established in 2012 o Is owned by a Kinh family o Was formerly supported by the Japan International Corporation Agency (JICA), SNV, followed by the ILO and UNESCO o Services include gardening experience and cooking classes o Currently in partnerships with 2–3 tour operators o Is a member of Thanh Toan's CBTE cooperative
Minh Tho Homestay	o Mai Hich Village, Hoa Binh Province o 5 km from Lac Village - a renowned and arguably unsuccessful CBT destination in Vietnam	o Is owned by a Thai[3] family o Initiated in 2011 with the support of COHED[4] o Has recently received marketing support from CBT Travel[5] o Offers homestay accommodation and other service packages (trekking, cultural performance, boating and biking)

[1] The Kinh people are the majority ethnic group of Vietnam.
[2] The CBTE cooperative is a form of community alliances that specialises in tourism. This co-op is a community institution consisting of member CBTEs and acting as a representative of the member entrepreneurs. Usually, a committee of selected members is responsible for the management of the co-op. In Vietnam, CBTE co-ops, similar to other communal cooperatives, are legally integrated into the over-arching Vietnam Cooperative Alliance, a non-profit organisation whose purpose is to support members through consulting, training and providing a voice for policy change.
[3] The Thai people are one of the minority ethnic groups of Vietnam.
[4] COHED: Centre for Community Health and Development.
[5] CBT Travel is a travel agency specialising in CBT products and services in Vietnam, self-labelled as a social enterprise. It initiated the 'franchised CBT' approach. Under this approach, CBT Travel facilitates CBT initiatives (mostly homestays) equipped with standard facilities and services to fulfil travellers' needs. The projects are then handed over to local entrepreneurs through franchising partnerships in which CBT Travel is responsible for sales, marketing and service quality control related to the projects. At the time of the investigation, CBT Travel supported Minh Tho Homestay with sales and marketing.

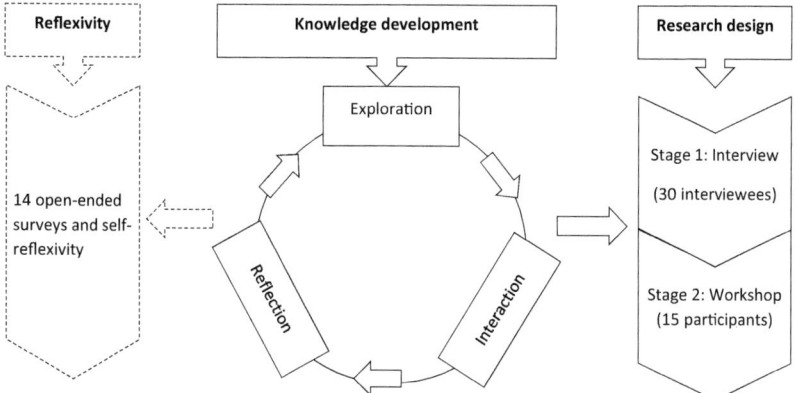

Figure 5.1 Knowledge development process included in the research design
Source: Adopted from PAR process of Kindon et al., 2007.

Research design

The research was designed in association with the knowledge development process framed by the three-stage PAR exemplar of Kindon *et al.* (2007). The knowledge development process began with a knowledge exploration stage, followed by knowledge interaction and then a knowledge reflection stage before proceeding to the next process. Along with this process, the research design consists of two stages. Figure 5.1 visually presents the linkages between the knowledge development process underpinned by PAR and the stages of the research design.

Stage 1 - Interviews

The perspectives of stakeholders regarding CBTE collaborative marketing alternatives and marketing proposals for CBTE sustainability were investigated. To address this objective, the interviewing method was adapted. Thirty key stakeholders directly involved in the three CBTEs were identified as interview respondents through the purposive sampling method and snowball techniques. Respondents were asked about their viewpoints related to who is included in and who facilitates a CBTE marketing network, the factors of success for CBTE collaborative marketing and how such an approach leads to the business sustainability of CBTEs. They were also asked about a proposal of marketing strategies oriented towards CBTE long-term success. The interviews were conducted face-to-face in public places at the convenience of the respondents in the period from November 2015 to January 2016.

Content analysis and narrative analysis were utilised to interpret the interview data. As a result, collaborative marketing alternatives for the business sustainability of CBTEs shaped by stakeholders from different perspectives and categories were elucidated. From these collaborative marketing alternatives, the incongruence of perspectives on CBTE collaborative marketing emerged.

Stage 2 - Workshop

At this stage, stakeholders with different perspectives communicated to explore other perspectives and obtain a shared understanding regarding a collaborative marketing approach for the business sustainability of CBTEs. Knowledge interaction occurred on a platform that was established using collective learning tools. In this study, a group discussion was selected to facilitate knowledge interactions (Barbour, 2008). Because of time and budget limitations, the group discussion took the form of a half-day workshop.

Major tasks involved in the workshop planning included setting the day, time, and venue of the workshop, arranging the incentives for workshop participants, inviting participants and maintaining their interest in attending the workshop. As the workshop involved multiple stakeholders from both the public and private sectors, a Friday morning was suggested as the most appropriate day. Additionally, the local community hall was deemed the best option to serve as the workshop's venue, given that the workshop was about CBTEs, collaborative marketing and business sustainability. Regarding the participation incentives, workshop participants were offered the option of pre-arranged one-night accommodation or an equivalent amount of cash to maximise their likelihood of attending. As a result, the workshop was held on the morning of 17 March 2017 in the community hall of Triem Tay Village, Quang Nam province, Vietnam.

Regarding the workshop participants, thirty interview respondents were targeted with an invitation email or phone call. The email or phone call specified the workshop objectives, i.e. sharing interview outcomes and discussing CBTE collaborative marketing issues that emerged from the interviews. Only five invitees out of 30 accepted the invitation. Three other participants suggested sending representatives to participate in the workshop. Considering a desired group discussion size between eight and twelve participants (Jennings, 2010; Krueger & Casey, 2014), the recommendations of participants who had accepted, and the need to have representatives of all categories of CBTE stakeholders, three additional participants were invited. Overall, 11 people accepted invitations to participate. Unexpectedly, four participants, who had heard about the workshop expressed interest in attending and were accepted because the workshop had the purpose of obtaining and sharing knowledge. Therefore, 15 people ultimately attended the workshop. Following the invitations, various activities were conducted to maintain the participants' interest in the workshop. In particular, a video briefly summarising the interview outcomes and introducing the workshop objectives was sent to participants in early March. An infographic detailing the collaborative marketing alternatives was sent to the participants one week prior to the day of the workshop.

The workshop was conducted in the form of roundtable discussions to facilitate a power-free environment for all voices to be raised and appreciated (Ross et al., 2015). Participants discussed the explanation behind diverse perspectives on CBTE collaborative marketing and negotiated with each other to finalise a common understanding of the topic. Participants were also provided with a proposed marketing strategy tailor-made for CBTE sustainability in Vietnam for a review and feedback.

Reflexivity

Participation in knowledge interactions during the research process is regarded as a learning opportunity that generates learning outcomes for all the participants involved (Berkes, 2009; Maiello et al., 2013). Thus, both the first author and the research participants were expected to learn relevant knowledge by 'trading-off' with other perspectives. Tools for evaluating the learning experience were identified through surveying the workshop participants and the self-reflexivity of the first author. At the end of the workshop, fourteen surveys were distributed to participants, with one participant leaving before the end of the workshop. Concomitantly, the first author recalled transformations in her research skills and accumulations of new insights owing to her entanglement throughout the research process. The first author's viewpoint had been nurtured within the academic world exclusively through her background education and working experience as a tourism lecturer from a developing country, and then flourished during the first year being immersed in Western schools of thoughts in her PhD tenure in Australia. The intellectual interactions had certainly shaped her assumptions relating to the research topic. However, in the first interview with a local authority, an incongruence emerged between her pre-assumptions and the interviewee's viewpoints. This difference of perspectives forced the first author to re-think the gap between these two sources of knowledge. Accordingly, the terms of 'constructivist paradigm' and 'knowledge co-production' came in to elucidate the research process, guide the data analysis and regulate the research outcomes (see further in Ngo et al., 2018a).

Lessons adhered to the engagement of multiple voices to co-generate knowledge in a knowledge co-production study

Along with the implementation of the research process, the principle of power-sharing was practised. Power-sharing is highlighted as one of the advantages of PAR (Liamputtong, 2013). In PAR, the power of legitimate knowledge holders, which is conventionally possessed by researchers, is decentralised to include research participants. Thus, PAR encourages empowerment and confidence among research participants to engage in the research process. Specifically, a knowledge co-production study, framed by PAR principles, advocates for the decolonisation of ways of knowing among co-researchers through their engagement in the research process (Chambers & Buzinde, 2015). However, translating this attribute into various stages of the research process is a challenge that has repeatedly been raised in knowledge co-production studies (Cornwall, 2004; Pohl et al., 2010).

Accordingly, in our study, various steps were taken to facilitate a sense of fair relations of power and knowledge among research stakeholders. First, in terms of data collection, the interview questions were designed and delivered flexibly to optimise the exploration of respondents' perspectives. They were structured in a basic manner with 'How do you think' or 'What are your ideas' questions supplemented by 'why' questions to clarify the respondents' viewpoints. Additionally, the interview questions were intentionally flexible to make them answerable within the boundaries of all respondents' perceptions. Likewise, in the workshop, the participants were distributed across three concurrent group discussions to ensure that knowledge interactions amplified the voices of marginalised knowledge holders (e.g. local entrepreneurs, community representatives). Additionally, open-ended survey questions at the end of the workshop encouraged participants to add any ideas that they had not shared during the workshop.

Second, regarding the data analysis, the assessment of interview data was independent of Western-driven knowledge. Furthermore, a narrative analysis was employed to explore insights from the respondents' perspectives following their own way of knowing. Third, the role of the principal researcher as a knowledge facilitator, rather than a knowledge disseminator, was reiterated via different channels of communication with each research participant (i.e. during interviews, in emails, in the video and the infographic sent to the workshop participants, and during the workshop). Finally, the research participants were allowed to learn about the research outputs through the video, infographic and presentation briefing illustrating the research findings that were delivered to them. Owing to these techniques, respondents' perspectives were elaborated and justified to engrave valid knowledge values on the topic (see Ngo et al., 2018a, 2018b, 2019).

Simultaneously, the trustworthiness of data was also taken into consideration. This trustworthiness is identified through four criteria, that is, credibility (internal validity), transferability (external validity), dependability (reliability) and confirmability (objectivity) (Shenton, 2004). The strategies recommended by Shenton (2004) to promote the trustworthiness of data in the qualitative research process were applied in our study. For instance, the first author was acquainted with the study context (Vietnam) beforehand. The field trip to undertake the interviews took over four months in order to interview 30 participants. The first author consistently implemented reflexivity during the data collection and analysis stages. Data triangulation was achieved as follows: the data collection was undertaken in two stages – interviews and a workshop; the informants were able to check the information after the interviews and workshop; document analysis of the current wealth of knowledge was conducted in conjunction with primary data analysis. The initial data analysis completed by the first author was reviewed by the other two authors (in the role of supervisors), academic fellows (through submitted manuscripts to peer-reviewed publications) and CBTE stakeholders (via the workshop).

Although this study has used the principles of the knowledge co-production approach, it fails to satisfy the attributes of authorship sharing and identification of knowledge holders. Given that the research participants were Vietnamese and their official language was Vietnamese, attribution of authorship was problematic and not addressed in this research. The Australian-based researchers were also confronted with language barriers, as they could not communicate directly with Vietnamese participants. As a compromise, research participants were added to the publications in the acknowledgements section. Development of further research methods that seek to facilitate the authorship of research participants of multiple languages within the co-constructed method is warranted.

Conclusion

This chapter elucidates the knowledge development process incorporated into a knowledge co-production study investigating a collaborative marketing approach for the business sustainability of CBTEs in less-developed countries. The knowledge development process is methodologically underpinned by the three-stage PAR framework developed by Kindon et al. (2007). The research paradigm presented in this chapter illustrates that PAR provides a framework for the attributes of a knowledge co-production approach to be embedded in the stages of the research process. Concurrently, this research paradigm indicates that value-based knowledge and social learning outcomes can be generated through the involvement of various research stakeholders.

Through depicting the research paradigm and highlighting the lessons drawn from the translation of the abovementioned methodological insights into the research process, this chapter can benefit readers with how to effectively articulate the principles of a knowledge co-production approach and PAR into a study framed by a constructivist approach.

References

Armitage, D., Berkes, F., Dale, A., Kocho-Schellenberg, E. and Patton, E. (2011) Co-management and the co-production of knowledge: Learning to adapt in Canada's Arctic. *Global Environmental Change* 21 (3), 995–1004.

Barbour, R. (2008) *Doing Focus Groups*. London: Sage.

Becken, S., Zammit, C. and Hendrikx, J. (2015) Developing climate change maps for tourism: Essential information or awareness raising? *Journal of Travel Research* 54 (4), 430–441.

Berkes, F. (2009) Evolution of co-management: Role of knowledge generation, bridging organisations and social learning. *Journal of Environmental Management* 90 (5), 1692–1702.

Carr, A., Ruhanen, L. and Whitford, M. (2016) Indigenous peoples and tourism: The challenges and opportunities for sustainable tourism. *Journal of Sustainable Tourism* 24 (8-9), 1–13.

Castleden, H., Morgan, V. and Lamb, C. (2012) 'I spent the first-year drinking tea': Exploring Canadian university researchers' perspectives on community-based participatory research involving Indigenous peoples. *The Canadian Geographer/Le Géographe Canadien* 56 (2), 160–179.

Chambers, D. and Buzinde, C. (2015) Tourism and decolonisation: Locating research and self. *Annals of Tourism Research* 51, 1–16.

Cole, S. (2006) Information and empowerment: The keys to achieving sustainable tourism. *Journal of Sustainable Tourism* 14 (6), 629–644.

Cornwall, A. (2004) Spaces for transformation? Reflections on issues of power and difference in participation in development. In S. Hickey and G. Mohan (eds) *Participation – From Tyranny to Transformation? Exploring New Approaches to Participation in Development* (pp. 75–91). London: Zed Books.

Dale, A. and Armitage, D. (2011) Marine mammal co-management in Canada's Arctic: Knowledge co-production for learning and adaptive capacity. *Marine Policy* 35 (4), 440–449.

Davidson-Hunt, I.J. and O'Flaherty, R.M. (2007) Researchers, indigenous peoples, and place-based learning communities. *Society and Natural Resources* 20 (4), 291–305.

Dixey, L.M. (2008) The unsustainability of community tourism donored projects: Lessons from Zambia. In A. Spenceley (ed.) *Responsible Tourism: Critical Issues for Conservation and Development* (pp. 323–342). London: Earthscan.

Dodds, R., Ali, A. and Galaski, K. (2016) Mobilizing knowledge: Determining key elements for success and pitfalls in developing community-based tourism. *Current Issues in Tourism* 21 (13), 1547–1568.

Dredge, D. and Jamal, T. (2015) Progress in tourism planning and policy: A post-structural perspective on knowledge production. *Tourism Management* 51, 285–297.

Espeso-Molinero, P., Carlisle, S. and Pastor-Alfonso, M.J. (2016) Knowledge dialogue through Indigenous tourism product design: A collaborative research process with the Lacandon of Chiapas, Mexico. *Journal of Sustainable Tourism* 24 (8-9), 1331–1349.

Font, X. and McCabe, S. (2017) Sustainability and marketing in tourism: Its contexts, paradoxes, approaches, challenges and potential. *Journal of Sustainable Tourism* 25 (7), 869–883.

Goodwin, H. and Santilli, R. (2009) Community-based tourism: A success? *ICRT Occasional Paper* 11 (1), 1–37.

Guba, E.G. (1990) *The Paradigm Dialog*. Newbury Park, CA: Sage Publications.

Guba, E.G. and Lincoln, Y.S. (1994) Competing paradigms in qualitative research. In N.K. Denzin and Y.S. Lincoln (eds) *Handbook of Qualitative Research* (pp. 105–117). Thousand Oaks, CA: Sage.

Hollinshead, K. (2006) The shift to constructivism in social inquiry: Some pointers for tourism studies. *Tourism Recreation Research* 31 (2), 43–58.

Holmes, A.P., Grimwood, B.S.R., King, L.J. and Lutsel, K.D.F.N. (2016) Creating an Indigenized visitor code of conduct: The development of Denesoline self-determination for sustainable tourism. *Journal of Sustainable Tourism* 24 (8-9), 1177–1193.

Idziak, W., Majewski, J. and Zmyślony, P. (2015) Community participation in sustainable rural tourism experience creation: A long-term appraisal and lessons from a thematic villages project in Poland. *Journal of Sustainable Tourism* 23 (8-9), 1341–1362.

Jamal, T. and Watt, E.M. (2011) Climate change pedagogy and performative action: Toward community-based destination governance. *Journal of Sustainable Tourism* 19 (4-5), 571–588.

Jennings, G. (2010) *Tourism Research* (2nd edn). Milton: John Wiley and Sons Australia Ltd.

Kindon, S., Pain, R. and Kesby, M. (2007) *Participatory Action Research Approaches and Methods: Connecting People, Participation and Place*. New York: Routledge.

Krueger, R.A. and Casey, M.A. (2014) *Focus Groups: A Practical Guide for Applied Research*. Thousand Oaks, CA: SAGE Publications.

Liamputtong, P. (2013) *Qualitative Research Methods*. Melbourne: Oxford University Press.

Maiello, A., Viegas, C.V., Frey, M. and Ribeiro, D. (2013) Public managers as catalysts of knowledge co-production? Investigating knowledge dynamics in local environmental policy. *Environmental Science and Policy* 27, 141–150.

Marshall, N., Viegas, C.V., Frey, M. and Ribeiro, J.L.D (2016) Advances in monitoring the human dimension of natural resource systems: An example from the Great Barrier Reef. *Environmental Research Letters* 11 (11), 141–150.

Mbaiwa, J.E., Stronza, A. and Kreuter, U. (2011) From collaboration to conservation: Insights from the Okavango Delta, Botswana. *Society and Natural Resources* 24 (4), 400–411.

Mielke, E.J.C. (2012) Community based tourism: Sustainability as a matter of result management. In G. Lohmann and D. Dredge (eds) *Tourism in Brazil* (pp. 30–43). London: Taylor & Francis.

Ngo, T., Hales, R. and Lohmann, G. (2018a) Social learning of co-researchers in a knowledge co-production study. Paper presented at *The Council for Australasian Tourism and Hospitality Education CAUTHE Conference*, Newcastle, Australia.

Ngo, T., Hales, R. and Lohmann, G. (2019) Collaborative marketing for the sustainable development of community-based tourism enterprises: Reconciliation of diverse perspectives. *Current Issues in Tourism* 22 (18), 2266–2283.

Ngo, T., Lohmann, G. and Hales, R. (2018b) Collaborative marketing for the sustainable development of community-based tourism enterprises: Voices from the field. *Journal of Sustainable Tourism* 26 (8), 1325–1343.

Nowotny, H. (2003) Democratising expertise and socially robust knowledge. *Science and Public Policy* 30 (3), 151–156.

Pohl, C., Rist, S., Zimmermann, A., Fry, P., Gurung, G.S., Schneider, F., Speranza, C.I., Kiteme, B., Boillat, S., Serrano, E., Hadorn, G.H. and Wiesmann, U. (2010) Researchers' roles in knowledge co-production: Experience from sustainability research in Kenya, Switzerland, Bolivia and Nepal. *Science and Public Policy* 37 (4), 267–281.

Reason, P. and Bradbury, H. (2008) *The Sage Handbook of Action Research: Participative Inquiry and Practice*. London: SAGE Publications.

Rocharungsat, P. (2008) Community based tourism in Asia. In G. Moscardo (ed.) *Building Community Capacity for Tourism Development*. Wallingford: CABI Publishing.

Ross, H., Shaw, S., Rissik, D., Cliffe, N., Chapman, S., Hounsell, V., Udy, J., Trinh, N.T. and Schoeman, J. (2015) A participatory systems approach to understanding climate adaptation needs. *Climatic Change* 129 (1-2), 27–42.

Shenton, A.K. (2004) Strategies for ensuring trustworthiness in qualitative research projects. *Education for Information* 22 (2), 63–75.

Torres-Delgado, A. and Saarinen, J. (2014) Using indicators to assess sustainable tourism development: A review. *Tourism Geographies* 16 (1), 31–47.

Ungar, P. and Strand, R. (2012) Inclusive protected area management in the Amazon: The importance of social networks over ecological knowledge. *Sustainability* 4 (12), 3260–3278.

Weiss, K., Hamann, M. and Marsh, H. (2013) Bridging knowledge: Understanding and applying indigenous and western scientific knowledge for marine wildlife management. *Society and Natural Resources* 26 (3), 285–302.

6 The Constructivist Paradigm and Phenomenological Qualitative Research Design

Justyna Pilarska

Introduction

As a set of assumptions, beliefs and models of conducting research, paradigms are fundamental (and primary) factors in the design of the inquiry. Not only do they orientate a researcher towards the (social) worlds to be investigated providing the appropriate 'tool' of getting insight, but equally they reflect the values that motivate a researcher in the undertaking. What motivated me personally in choosing constructivism and a phenomenological qualitative research design was, from my perspective, their deeply humanistic and respectful approach to the researched and their cultural as well as social and psychological realities, followed by their dialogical, interpretive nature and interactive dynamics. Although constructivism is not a homogenous paradigm (to recall, for instance, radical constructivism, social constructivism, constructionism) (Hacking, 1999), all constructivist approaches acknowledge that there are no universal 'truths' or valid categories for human experience; hence knowledge is the production of social and personal processes of meaning making.

Such a stance highlights the principle that people construct their own understanding and knowledge of the world through experiencing things and reflecting on those experiences (Honebein, 1996), and this is the assumption I adopted for my own research design.

Constructivism evolved from discontent with traditional ways of acquiring knowledge. It contended that new knowledge is constructed. Thus, it is a concept formed as opposed to the philosophical rationalism, epistemological fundamentalism, essentialism, objectivism and realism. As Yvonna Lincoln implies, the constructivist paradigm does not make its heart of the matter 'the abstraction (reduction) or the approximation (...) of a single reality but the presentation of multiple, holistic, competing and often conflictual realities (including the inquirer's)' (1990: 38). While working with the constructivist paradigm I realised and acknowledged that the social and cultural realities to be investigated are subjective, so the research process within the constructivist paradigm relies heavily on my own (the researcher's) understanding of human beings at a deeper level, and by exploring the phenomena more intensively. Therefore, because they constitute the building blocks of a socially constructed reality, social and cultural facts are the essence of constructivism. Such an approach to the research can be facilitated by phenomenology, which – as a Husserlian perspective – is a method of inquiry that embraces not only the epistemological approach to knowing, but a manner of understanding. According to Husserl (1977), intellectual involvement in interpretations and meaning making are inseparable acts of understanding the lived world of human beings at a conscious level. Consequently, the description of the invariant aspects of phenomena as they appear to consciousness (or, as constructivists would say, subjectively) is feasible. Along these lines, the Husserlian phenomenological research method focuses on exploring (seeking) the multiple realities, expressed in the form of worlds made of interconnected, subjectively lived experiences (Crotty, 1998).

When conducting field research related to cultural matters that involve cross-cultural identities, intercultural communication, functioning of the ethnic or religious minorities (or other issues concerning cultural diversity), adapting constructivism alongside phenomenology helped me reach my research objectives in an ethical, thoughtful and inspiring manner. This is true also for other qualitative methods and paradigms; however, it was phenomenology and constructivism that allowed my being-in-the-world as a researcher to adopt a respectful presence. The reflections below align with the tenets of the constructivist paradigm and phenomenology, entailing an explanation of how I applied the paradigm and methodology in the specific context of my research, followed by reflexivity regarding their benefits and challenges in the research practice.

Review of the Paradigm: Constructivism and Phenomenology in a Research Design

Social constructivism, often linked to interpretivism (cf. Mertnens, 1998), is an orientation applied in qualitative research that originated from Mannheim's philosophical assumptions as explained by Berger and Luckmann (1967). Ontologically, constructivism implies that realities are socially constructed and are hence ungoverned by universal laws (cf. Lincoln & Guba, 2000; Neumann, 2000; Schwandt, 2007). Such realities are therefore dependent for their form and content on the persons who hold them. The multiple realities (mentally constructed by the individuals) are subject to local and specific truths, hence their relative character. Therefore, the assumption that individuals assign subjective meaning to the experience within their worlds led me to the recognition of the fact that the researched are the co-producers of the knowledge that comes within their own selves, and is expressed through interaction with me as the researcher.

Epistemologically, it is a subject- and object-interrelated view. On that account, the methodology is hermeneutic (explanatory) and dialectic (pertaining to the logical argumentation) in nature, as it is interpretive (hermeneutic concerning interpretation) and involves dialogic techniques of reasoning by dialogue, as an intellectual exchange of ideas. In this regard, hermeneutics as the methodology of interpretation helped me in solving problems of interpretation of human actions, whereas the dialectic nature of the paradigm implied the Socratic dialectical method of cross-examination, as I became part of the reality that was being researched. Consequently, the research findings became a creation of the inquiry process itself, rather than a collection of external, already existing systems. This provides evidence that every social phenomenon can be studied as the result of a construction process (Hacking, 1999).

What seemed intriguing for me as a researcher, was that the constructivist paradigm questions a concept of 'reality' as something that is 'objectively given', instead focusing on the construction processes implied in the creation and establishment of 'reality'. This explains why constructivist perspectives imply ontological considerations, as embedded in (social) practice itself. Such an approach determines methodologically the view on the nature of knowledge. The latter is, according to the constructivist paradigm, constitutively conventional; that is, it is a cultural construct (Law & Lodge, 1984). Consequently, it can only exist within the human mind and does not have to match any real-world reality. Such an approach to knowledge had for me a specific impact on the notion of truth, which in this stance was to be recognised as not absolute or immutable, and understood as the best informed and most sophisticated construction (Schwandt, 1994).

Constructivism relies on the principles of social as well as cultural interpretation; that is, constructivist researchers question traditionally held ideas on the possibility of acquiring objective knowledge concerning the real world. The latter is assumed to be subject to cognition via the procedure of reproducing its structure in the process of thinking. Thus, the traditional (realistic and objective) epistemological determination is denied, because in the constructivist view knowledge is not about revealing the 'truth' about reality, but about constructing (creating) such individual, personal 'truth' about the world (Devitt, 1997). It is a natural consequence of acknowledging the fact that the world, as an object of cognition, does not exist objectively; that is, beyond sociocultural interpretations. Therefore, if everything we treat as the object of our cognitive interest is recognised through given sociocultural 'filters', it is assumed that the objective truth recognised scientifically does not exist. As a result, it can be considered only metaphysically as (the previously recalled) worldview. The origins of such a standpoint can be found in Immanuel Kant (Gardner, 1999), who reversed the relation between the cognitive subject and object, claiming the world presents itself to a human by the agency of human cognitive skills, namely sensory inspection and thinking. Correspondingly, the world adapts to the cognitive possibilities of a human who, becoming an active participant of the process, 'creates' the object under cognition (Rockmore, 2008).

As far as methodological design is concerned, I recognised the fact that most constructivists do not begin a research project with a theory but they 'generate or inductively develop a theory or pattern of meanings' throughout the research process (Creswell, 2003: 9). It is understood that the research methodologies for the constructivist philosophical paradigm include: narrative study, case study, ethnographic study, grounded theory, descriptive study and phenomenological study (e.g. Kivunja & Kuyini, 2017; Mackenzie & Knipe, 2006; Manning, 1997). During data analysis, the researcher constructs an image of the phenomenon that takes shape as s/he collects and examines the parts of the data collected (Bogdan & Biklen, 1992). For that reason, phenomenological 'bracketing' of one's own biases is of paramount importance and I made it my priority in the process of self-reflection, bearing in mind that from a constructivist perspective, paradoxically, 'it is not possible to be completely free of bias' (Strauss & Corbin, 1998: 97).

Methodologically, and adopting Creswell's (2003) principles, constructivism as a paradigm can be expressed by the following tenets:

- social and cultural construction of the reality is reflected in the research findings through multiple interpretations and comprehensive understanding;
- understanding the influence of social behaviour or the attitude of individuals in a particular community takes place by considering a wide range of views on the given phenomenon, as presented by the research participants;
- comprehension of the phenomenon being studied is done by seeking to understand how individuals make sense of their everyday lives in their natural settings (e.g. in the local communities);
- comprehending the practices of a group or society and its implications towards their attitudes assigns paramount (and exclusive) importance to the subjective meanings that can be socially or historically conditioned (these subjective meanings are generated upon interactions and through the impact of historical and cultural norms on the individuals);
- the constructivist researchers are focused on the specific context that people live in, act and interact; that is, the key role of life histories or life stories of personalities in communities, the oral history of a clan, ethnic society, minority groups, etc.

As in the constructivist paradigm, the researcher seeks in-depth understanding, I was obliged (and had to be prepared and genuinely willing) to interact with those being researched. Such a stance implies that the findings are the creation of the interactive process, hence subjectivity is an inevitable part of the design. The latter, as a result, is flexible and emergent (Crotty, 1998).

From the starting point, the constructivist inquiry is subjective and individual, and it corresponds with phenomenology, so the two can be matched and successfully applied in the research design. Similar to the assumption of the constructivist paradigm, and in addition to its descriptive nature, phenomenology gave me a wider meaning to the lived experiences under study.

Since phenomenology as a methodology is more elaborate and complex (sophisticated) than the tenets of the constructivist paradigm, I would like to draw more attention to the peculiarities and details of phenomenology, which constitutes a part of the constructivist (interpretivist) paradigm, and concerns both philology and methodology (Spiegelberg, 1969). As a research methodology, phenomenology is tightly linked to the disciplinary field of philosophical inquiry of the same name. Although methodologically it orientates the researcher towards in-depth exploration of the phenomenon under study through consciousness (Creswell, 2007), some scholars (cf. Spiegelberg, 1969) argue that there is no one style of phenomenology as the researcher can apply different research approaches, and a 'consensual, univocal interpretation of phenomenology is hard to find' (Giorgi & Giorgi, 2003: 23).

Albeit historically, phenomenology can be traced back to the epoch of Plato, Socrates and Aristotle (Fochtman, 2008) where it was referred to as a philosophy of human being, the German philosopher Edmund Husserl is considered to have established phenomenology as an approach to study lived experiences of human beings at the conscious level of understanding (Fochtman, 2008; Wojnar & Swanson, 2007). Then, one of Husserl's students, Martin Heidegger, developed his own method of phenomenology, e.g. an interpretive-hermeneutic one, entailing key phenomenological ideas such as everyday ordinariness, *Dasein*, being in the world, being with, and temporality (Heidegger, 1996). Methodologically, there is a distinction between hermeneutics and phenomenology. The first one can be applied in a research agenda involving interpreting texts to explore lived experiences, engaging with research questions of a wide range of social and humanistic issues (Landgridge, 2008). It can concern, for instance, women's experiences of being mothers, as interpreted through blogs, social media accounts and tweets. The Heideggerian view of interpretive-hermeneutic phenomenology gives a wider meaning to the lived experiences under study. The Husserl's transcendental phenomenology, which is the one I applied for my own research project, concentrates on people's meaning of a lived experience of a concept or phenomenon. Consequently, the purpose of such research is to describe the essence and the nature of experiencing phenomena. Hence, the research question in this regard would be, analogically, 'What is the experience of being a mother?'.

Epistemologically, phenomenological approaches are based on a paradigm of constructive, personal knowledge and subjectivity, and emphasise the importance of personal perspective and interpretation. From the methodological view, phenomenology is not a closed philosophical system, but rather a method of describing and understanding phenomena; that is, what emerges in the study of a phenomenon is reflective of what presents itself (Husserl, 1970). Hence, the act of consciousness and the object of consciousness are intentionally related in the study of phenomena. Husserl maintained that 'intuition' is therefore essential in describing whatever presents itself. Moreover, he preferred using intuition over deduction. Interestingly, in the volume on Edmund Husserl, edited by Robert Sokolowski, Richard Cobb-Stevens says 'as early as 1894, he contended that intuitive consciousness reaches out beyond sensory impressions, and grasps the intended object itself' (Cobb-Stevens, 2018: 55). As a form of knowing, intuition constitutes a key feature of the researcher, followed by the need to bracket one's own previous knowledge (entailing aspects such as prejudices, assumptions and biases). These provided me with a strong indication on how to conceptualise my research design ontologically. The concept of 'bracketing' (Gearing, 2004) involves two negative procedures that Husserl disclosed, i.e. the *epoché of the natural sciences* (return from theories to the things themselves, i.e. avoiding explanations), and the *epoché of the natural attitude,* that is phenomenological reduction. The latter implies becoming unaware of the presumptions and presupposition that researchers keep in their mind, and concentrating on the original phenomenon the way it manifests itself to the research participants rather than how the researcher is involved with the phenomenon (Qutoshi, 2018: 218). Correspondingly, the two main positive procedures Husserl developed are *intentional* and *eidetic analysis* (Qutoshi, 2018: 218).

Epoché embraces four planes, and each 'between' constitutes another form of reduction, i.e. the theoretical world, the world of everyday life or natural approach, phenomenological approach and the examination of the essence, and last but not least, the transcendental subjectivity. Taking that into account, Moustakas (1994) explains:

- the first *epoché* is the shift from the theoretical world to the world of everyday life towards one which takes a natural approach as a suspension of judgments about the existence or non-existence of the external world – this helped me focus exclusively on the phenomena;
- the second *epoché*, referred to as phenomenological reduction that leads from a natural attitude to phenomenological attitude, where both of worlds, i.e. the theoretical one and the one of everyday life, are 'suspended' – thanks to this I could examine the phenomenon taking into consideration its horizon, perspective and time;

- eidetic reduction originating from phenomenological attitude is the third *epoché* focusing on the disclosure of the essence of the phenomenon, which helped me learn which characteristics are necessary for an object to be it without being something else;
- transcendental reduction can be defined as the fourth *epoché* which is about exploring the core of the phenomenon as a reduction to 'pure' consciousness, i.e. to be intentionally purified of all psychological, all 'worldly' interpretations and to describe the phenomenon simply as it presents itself.

Consequently, the range of phenomenological reductions encompasses the following levels:

- from suspending the theoretical world to the world of everyday life;
- from natural approach (natural attitude) to the phenomenological attitude;
- from phenomenological attitude to the examination of the essence;
- from phenomenological and eidetic reduction to the world constituted in the *transcendental I* (Ablewicz, 1994; Moustakas, 1994).

As a methodology, phenomenology has been profoundly examined by Clark Moustakas (1994) in his study *Phenomenological Research Methods*. I found his book helped me to better understand phenomenology. He defined and explained the key methodological terms, including phenomenological reduction (bracketing the topic or question), horizontalisation, individual textural description (i.e. descriptions with a context in experiencing), as well as composite textural description and imaginative variation. As a qualitative strategy, phenomenology recognises the sense of human experiences towards a given phenomenon on the basis of the meaning that is assigned to such phenomena (Nieswiadomy, 1993). As a researcher I needed to bracket, or put aside, own experiences to fully understand (and immerse in) the experiences of the participants (cf. Nieswiadomy, 1993). Therefore, to obtain as truthful a description as possible in my phenomenological project, I attentively followed qualitative approaches to inquiry. Moreover, as a qualitative researcher I believed that human experiences are not approachable through quantitative approaches, and so, I had to acknowledge the interpretive nature of knowing (Moustakas, 1994). For this reason, phenomenological research has overlaps with other essentially qualitative approaches including ethnography, hermeneutics and symbolic interactionism (Lester, 1999), for phenomenological research seeks essentially to describe rather than explain, and to start from a perspective free from hypotheses or preconceptions (Husserl, 1970). The focus in such design is on the wholeness of experience rather than solely on its objects or parts, as the main area of investigation concerns meanings and essences of experience, as in other qualitative approaches, rather than measurements and explanations

(Moustakas, 1994). As Moustakas explains, 'what appears in consciousness is an absolute reality while what appears to the world is a product of learning' (1994: 27). Methodologically, the premises of phenomenology in the research process which I followed imply the following:

- search for the essence – *eidos* – within the three aspects of the phenomenon, i.e. objective (phenomenon of something), subjective (phenomenon for someone) and horizontal (phenomenon remaining in a relation with other phenomena);
- perceptive acts – when referring to Husserl, as inspired by Descartes' belief, Moustakas claims that 'perception of the reality of an object is dependent on a subject' (1994: 27);
- phenomenological reduction – *epoché* as a reduction – i.e. exclusion and bracketing the researcher's approach to the world and his/her knowledge about the world (Husserl, 1977: 145);
- intentionality of the awareness in conformity with the assumption that the whole world – spatial and temporal – is an intentional existence, hence the place of the realistically existing world is 'replaced' by the world emerging within our awareness (*noema* as an object or content of a thought, judgment or perception);
- constitutions of a meaning rely on the assumption that the objective meaning, and the intentional activity are inseparable as intentionality is what leads to the awareness of the matters;
- *transcendental I* – understood as the necessity of the awareness of the subject's identity with the own self, with transcendental reduction facilitating the understanding that the awareness is transcendental and capable of constituting the meaning (Smith, 1993).

Summing up, phenomenological, qualitative proceedings within the constructivist paradigm put stress on the process of description and understanding that must take place so that phenomenological reduction can occur. As a result, the research focuses on processes of constructing (or reconstructing) meanings in a given cultural or social context without any presumptions concerning universality, focusing on the individual, and lived experiences. The research methods entail qualitative interviews, participant observation, action research, focus groups, analysis of diaries and other personal texts. Conversely, phenomenological premises of the research require to take into consideration the key tenets, i.e. engagement in the process of *epoché* as a manner of creating an appropriate atmosphere and contact, as well as 'bracketing' the question and conducting research using qualitative research methods, in order to obtain the descriptions of the experience. Through this approach, the researcher is able to reconstruct the essence of an experience referring exclusively to the participants (cf. Riemen, 1986), and this is what I was able to achieve and reconstruct by applying constructivism and phenomenology.

Application of the Paradigm

The topic of my doctoral dissertation concerned the process of conceptualising a multidimensional, cross-cultural identity in a culturally diverse community, exemplified by Bosnia-Herzegovina. The theoretical framework embraced such concepts as a culturally diverse society, Bosnian identity, axiology of the borderland, cultural transmission, the everyday life (corresponding to Heidegger's everyday ordinariness), cross-cultural learning and intercultural education. The theoretical objective of the project was to distinguish educational aspects of conceptualising multidimensional Bosnian identity, given the priorities of intercultural (informal) education. This objective was linked to the description of the social and cultural reality of Bosnia-Herzegovina, that can, in turn, provide a foundation for working out a new theoretical model of intercultural education related to explaining the essential indicators of the (inter)cultural dynamics. Subsequently, the project recognises the realities experienced (the lived experience) by the members of a culturally diverse community, as well as its implications, given conceptualisation of a multidimensional cultural identity. The research project was idiographic in nature. The practical objective concerned the recommendations contributing to the educational and cross-cultural discourse with the distinction of activities focusing on empowering multidimensional identities and strengthening the intercultural discourse.

The purpose of the research project was the analysis and categorisation of the conditions of the conceptualisation of identity in a culturally diverse environment (cultural borderland) exemplified by Bosnia-Herzegovina, encompassing the theoretical reflection within the discourse of intercultural education. A preliminary review of library findings in Bosnia-Herzegovina was focused on the exploration and analyses of the works of intercultural pedagogues and sociologists from the Bosnian cultural domain, concerning the Bosnian culture and its idiographic, emic uniqueness. During this time, I also undertook a critical discussion on the issue with regards to the Polish literature. Conducting a comprehensive review of the professional and research literature was feasible due to my good command of the local languages (Bosnian/Serbian/Croatian), which also allowed me to become an understanding partner while carrying out qualitative research in accordance with the premise of immersing myself into the cultural setting of the object (and subjects) of the research. The two main areas of research inquiry embraced the process of conceptualising the identity, and the impact of the cultural diversity on the process of conceptualising the identity. Thus, the research project tackled the influence of the cultural environment on the process of establishing ontological and axiological being. It acknowledged an interdisciplinary perception of the identity; that is, cultural, social, pedagogical and psychological perspective and narratives. The project revealed a number of local, idiographic peculiarities related to the multidimensional, multiple identities that members of a cultural borderland hold. Accomplishing the project enabled the linking of the theoretical, intercultural discourse with the practical dimension, locating the theoretical analysis in the current area of postmodern implications of the pedagogical reality, i.e. inter- and multiculturalism.

The research questions concerned the following:

- the process of conceptualising Bosnian, diverse identity, as taking place in a culturally diverse setting (cultural borderland);
- the role of the everyday life experiences at a cultural junction in the process of conceptualising the multidimensional identity.

Adopting the constructivist worldview and phenomenology as methodology with qualitative inquiry in this research project, I used multiple methods including interviews, observations and photography (visual anthropology). The constructivist paradigm provided me with the framework for designing my project in line with the research questions, whereas phenomenology allowed me to concentrate on the in-depth understanding of the phenomenon (the identity, the sense of being Bosnian), embedded within research participant's views and perspectives. Since the reporting of findings in phenomenological studies within a constructivist paradigm needs to be focused on detailed descriptions of the phenomena, my interview questions entailed multiple aspects of the experienced phenomenon (of being Bosnian). In-depth face-to-face interviews were of paramount importance to unearth the participants' perspectives and their experiences, whereas semi-structured interviews helped me maintain focus and direction. Accordingly, I prepared open-ended questions in a semi-structured format to help me clarify aspects during the interviews. The open-ended questions I asked were helpful in gathering a wide range of responses (cf. Creswell & Poth, 2017). It also served the purpose to build rapport with participants (cf. Lichtman, 2011; Merriam & Tisdell, 2016). The semi-structured format of the interviews (topical-guided) helped me maintain the focus of the study. They also suited the 'dynamics of the situation' (cf. Cohen *et al.*, 2007) between the interviewer and the interviewee. I was mindful that in phenomenology the questions should be broad, i.e. without any reference to the literature or existing classifications. Thus, my research questions tackled two key aspects: (a) what is it that the participants experience ('What does it mean for you to be Bosnian?') and (b) what is the context, i.e. the situation in which they experience this ('When do you realise you are Bosnian the most?') (cf. Nieswiadomy, 1993).

In general, the phenomenological method of inquiry within the constructivist paradigm demands the use of multiple instruments to gather data on participants' experiences, interpretations and understandings (Landgridge, 2008). As the objective of phenomenological inquiry is to look closely at the phenomena under study to explore the complex world of lived experiences from the participants' point of views, apart from the interviews, I applied participant observation and the visual method of taking photos to illustrate how the research participants' Bosnian identity is expressed in their daily lives, bearing in mind that, as Lichtman (2006: 8) points out 'the main purpose of qualitative research – whatever kind – is to provide an in-depth description and understanding of the human experience, human phenomena, human interaction, or human discourse'. Since phenomenology as a method relies on personal lived experiences (cf. van Manen, 1990) and interpretations (cf. Gadamer, 1997), the research participants are to meet one fundamental criterion: the selected interviewees must have experience with the phenomenon being studied.

I would advise purposive sampling in the data gathering process, as this is a qualitative sampling method that helps the researchers intentionally select individuals and sites to learn or understand the central phenomenon (Creswell, 2012). During my data collection period, I spent three months in Bosnia-Herzegovina conducting the research in various, numerous culturally diverse locations within Republika Srpska and the Muslim-Croat Federation. Primarily, I put myself through the process of bracketing, which involved aware practice to set aside my own experiences, biases, preconceived notions so as to understand how the phenomenon of being Bosnian appears to participants, instead of how it is perceived by me as a researcher. The selection of the research participants followed the key principle of embracing those who have had a lived experience with the concept or phenomenon of interest (being Bosnian). I continued with the in-depth individual interviews (and observation), until saturation took place. Observation concerned the elements of intercultural communication that expressed its symbolic, non-verbal manifestation; for example exclusive versus inclusive communication, daily cross-cultural social rituals in the public space, cultural codes of the body language. Following the initial stages of the field research, i.e. ensuring confidentiality, agreeing to a place and time convenient to the interviewees, and obtaining appropriate permission to record and publish the collected data, I conducted 30 phenomenological research interviews, which I correctly believed would ensure data saturation. It was the process of eidetic disclosure of what is invariant in each participant that gave me a sense of data saturation. The duration of the interview was between 2 and 3 hours so that the participants were not rushed and there was time and space for establishing a trustful, cooperative and amicable atmosphere. The interviews took place in different settings, sometimes at homes, or at public spaces such as parks or cafes. At all the premises, during the audio recording the interviews were accompanied by important cultural rituals to the Bosnian tradition, i.e. having small talk, drinking coffee, sharing meals and smoking cigarettes.

As I mentioned, the very first step of my research was to engage in the *epoché* process as a way of creating an atmosphere and rapport for conducting the interview. After bracketing the question and conducting the qualitative research interviews to obtain descriptions of the experience (of being Bosnian), I intuitively-reflectively integrated the composite textural and composite structural descriptions to develop a synthesis of the meanings and essences of the phenomenon or experience. By understanding the phenomena of being one's own self culturally, socially and psychologically, it also helped me explore my own (cultural) nature, bringing a transformation at a personal level. By doing so, phenomenology indeed facilitates the critical reflection for the researcher, who can become more thoughtful and attentive in understanding social and cultural practices.

As phenomenological research is a personal and at time intimate experience for the researcher, I would highly recommend that the same person (the researcher) executes the interview and the analysis. In analysing the gathered (fairly disorganised) data, the very first step is to read through the audio-recorded interview transcripts and get a feel for what is being said and making initial attempts at identifying key themes and issues in each text (Creswell, 2003). This can be facilitated with the aid of a mind-map. Respectively, a list of key themes can be used as a set of points to interrogate the texts, which are then summarised ('what is this participant saying about...'). In organising, analysing and synthesising the data I followed the modified Stevick–Colaizzi–Keen method which involves the synthesis of composite textural and composite structural descriptions (see Table 6.1). At all the stages it is important that the researcher looks for the participant's point of view on the aspects he or she refers to in the interview and maintains *epoché*, refraining from judging, labelling or criticising.

Summing up, I organised and analysed the data to facilitate the development of individual textural and structural descriptions, a composite textural description, a composite structural description and a synthesis of textural and structural meanings and essences. It is worth stressing that the data collection and meaning making in phenomenological research take place simultaneously. My purpose was to illuminate the specific experience to identify the phenomena (being a Bosnian) that is perceived by the participants (Bosnians) under particular circumstances. This also helped me, as the researcher, to be subjective and to gain personal knowledge in perceiving and interpreting the phenomena from the research participants' point of view.

As shown in Table 6.1, I applied horizontalisation which extracts significant statements from the transcripts to describe elements of experiencing the phenomenon. These significant statements encompass sentences and quotes that describe how the participants experience the phenomenon. I placed similar significant statements into 'clusters of meaning' with different themes of the participants' experience with the phenomena (*being Bosnian as a role, being Bosnian as a symbol, being Bosnian as a 'product' of intercultural interaction*). This was followed by textural

Table 6.1 The researcher's steps and activities in the modified Stevick–Colaizzi–Keen method as described by Moustakas (1994)

Step one	Step two	Step three	Step four	Step five	Step six
Researcher's epoché	Transcendental-phenomenological reduction	Imaginative variation (from different vantage points)	Synthesise	Repeat	Combine
Activities to step one	Activities to step two	Activities to step three	Activities to step four	Activities to step five	Activities to step six
-Bracket the question -Set aside judgment and prejudices -View the phenomenon with a fresh eye	-Consider the phenomenon with an open mind and from various perspectives -Identify units of meaning/ segments (invariant horizons) -Horizontalisation: each statement has equal value -Individual and composite textural description of the phenomenon: descriptive integration of the invariant textural constituents and theme of each research participant -Composite textural description: integration of all of the individual textural descriptions into a group of textural description	-Construct the structural epitome of the experience from the textural description -Individual structural description: integrating the structural qualities and themes into an individual structural description -Composite structural description: integration of all the individual structural descriptions into a group structural description of the experience	Intuitively and reflectively integrating the combined textural and composite structural descriptions	The process is followed for each participant until data saturation is achieved	The textural and structural descriptions to form a textural-structural essence of the experience (a composite description representing the essence of the experience)

Source: Adapted to own study based on Moustakas (1994).

description of participants' experiences as provided in the answers to the questions (what they experience). By contrast, the structural description (imaginative variation) was provided with the context and setting that revealed how the participants experienced the phenomenon. Then, I used the textural and structural descriptions to write the essence of the phenomenon. This is a long passage of text which gives the reader an understanding of what it would be like to experience the phenomenon, such as:

> '…it's not easy to answer such questions, because we usually don't think about who we are, it's good to think about who I am…well I am Bosnian because I speak Bosnian, I know Bosnian tradition, I was born in Bosnia and well, it means to love your country despite the rotten system, corruption, it means to love what you see when you wake up, your neighbourhood, your kids, well, it's my daily life, you know, that speaks for my Bosnianhood' [excerpt from the audio-recorded interview].

> 'Well…to be Bosnian, well it's not individual it's about us all […] I mean…I am Bosnian because I live in Bosnia and I coexist with Bosnians I guess… to be Bosnian means to be frank and open to others, no matter the nationality or religion, spontaneous and with vivid imagination, to have the soul; to be able to find solution in each situation; it means we, as Bosnians, we know how to relax, we are not afraid to experiment, we don't fret the new; if we were afraid of it our entire history would have looked differently and then perhaps indeed we would have killed each other' [excerpt from the audio-recorded interview].

To achieve the eidetic descriptions, I applied the following stages:

- I listened to each interview recording as many times as necessary to recall details so that the eidetic process could rely on the participant's important experiences and how he or she experienced them.
- I transcribed the interviews as an indirect discourse (in Bosnian and then in Polish), reporting on important experiences of the participants and how they were experienced (see Table 6.2 for details).
- I synthesised the important experiences and meanings that participants expressed. Upon establishing a summary of the findings based on the major themes that emerged from the data, I added direct quotations of the research participants to the major themes. The interpreting took place according to the Husserlian phenomenology through descriptions rather explanation (Husserl, 1977).
- I synthesised each participant's experience and meanings verifying their eidetic invariance.

Taking the above into account, the final conclusions of my research project entailed (among others) the following:

- The pluralistic everyday life reality is dynamised by the constant encounter with the other. The latter, in turn, enriches identities of the Bosnians with culturally diverse orientation of their self.

Table 6.2 Composite descriptions representing the essence of the experience of the interviewees

Bosnian individual identity	Bosnian social identity	Bosnian cultural identity
Who am I?	*How to be?*	*How to participate?*
• An open and aware individual • Self-conception on the basis of cultural code • Authenticity, integrity, autonomy, self-recognition	• Cognitive, emotional and behavioural openness • Combination of the individual and the collective in a culturally diverse setting • Axiological polymorphism	• The awareness of identity-creating impact of culture on the community • Creative and dynamic engagement in cultural creation • Active participation • Being aware of values • Emotional approach to values

Source: Author.

- The 'multidimensional identity' acknowledges the perception and interpretation of cultural diversity as the key component of the process of conceptualising one's own self, and becoming more self-aware culturally, socially and psychologically.
- The cultural transmission is open, dynamic and interactive.
- The Bosnians utilise daily experiences in the praxis of creating a culture of dialogue and enriching their identities with the sense of diversity.

In view of the above, I found that phenomenological methods within the constructivist paradigm are particularly effective at bringing to the fore the experiences and perceptions of individuals from their own perspectives, and therefore at challenging structural or normative assumptions. Phenomenology is an approach to educate our own vision, to define our position, to broaden how we see the world around us and to study lived experiences at a deeper level. It is the most appropriate choice if a researcher wants to employ a narrative, cooperative (participants as the source and co-producers of knowledge) manner of describing a given phenomenon to the wider audience through the local views of the participants.

Reflection on the Use of the Paradigm

Constructivism is an interpretive approach which puts strong emphasis on the meaning-making activity of the individual mind. Likewise, it is about approaching 'the object in a radical spirit of openness to its potential for new and richer meaning. It is an invitation to reinterpretation' (Crotty, 1998: 51). Similarly, phenomenology requires the researcher to approach the subject (or object) in an open-minded, bias-free manner. Since phenomenology entails suspension, bracketing and reduction of the researcher's preconceived notions and prejudices, it empowers constructivist methodology that is equally dialectical and relative. Blending these two concepts allowed me to design such research that reached the personal, subjective meanings assigned by the interviewees to their own concept of identity. Adopting a phenomenological methodology and following the constructivist principles of a facilitator of multi-voice reconstruction, I was able to 'decontextualise' myself in this culturally diverse setting. As a result, it ensured a 'fresh', undisturbed approach to the phenomena and interactions I was part of during my research.

Constructivism was the best option for my research design as it implies trustworthiness and authenticity as the key ethical principles of conduct, which in turn allowed me to meet the phenomenological objective of the inquiry, i.e. to describe things as 'they are', by understanding meanings and essences in the light of intuition and self-reflection. I chose phenomenology as it relies on individuals with their particular biographies and future perspectives as they share some common context. On the other hand, thanks to its subjectivist epistemology, constructivism allowed me to discover the phenomenon of identity, for the self is seen as something constructed through individual interaction and perception in an intersubjective world.

As qualitative research guided by the constructivist philosophy, my project enhanced descriptions of the experience (the 'subjective reality of being Bosnian') which were accessible via first-person accounts. This, in turn, outlined my research frameworks as fluid, for I operated within structures that are fluid (and intangible). I found that the most important assets of applying phenomenology within the constructivist paradigm is that it helps the researcher (as it helped me) move beyond one's own ego and methodically reveal the transcendent intersubjectivity of the phenomena present in the other's culture (multicultural Bosnia-Herzegovina), and society (Bosnian Muslims, Serbs and Croats). It also provides premises for research awareness of the value of Heideggerian everyday life, where individual and collective existences take place. Thus, phenomenology endorsed my focus on the context-bound, contextualised narrative data, with the emerging insights grounded in participants' experiences. Another benefit of applying phenomenology within the constructivist paradigm is that it avoids making pre-assumptions which often reify the subjects, concurrently giving space to concentrating on the specific subject of research (in this case Bosnian identity).

In terms of my own experience, the challenges related to research design utilising constructivism and phenomenology, are like any other qualitative research method, where the subjectivity of the gathered data is accentuated and entails the risk of making data too generalised or inaccurate because of biases towards the researcher's subjectivisms. Therefore, a researcher must practice their own interpersonal (soft) and intrapersonal skills of mindfulness, self-awareness, self-criticism, open-mindedness, dialogical communication and active listening skills. Analysing and mining data can also be challenging as it is a time-consuming, subjective effort. Nonetheless, phenomenology allowed me to reflect on my own self (as a researcher), i.e. to consciously experience the self as both researcher and learner of the new reality. Another significant advantage of phenomenology (specifically the Husserl's transcendental phenomenology) is that it puts emphasis on the necessity to consider an experience (e.g. being Bosnian) in its singularity not collectivity, i.e. in itself, and for itself. Moreover, the research process is iterative; that is, data collection and research questions are adjusted according to what is learned. This, in turn, enhances flexibility, creativity and innovativeness that can all contribute to the quality of the research design. Conversely of course such ephemeral and fluid atmosphere can trigger stress and anxiety that a researcher might experience with relation to the uncertainty that unfolds. Nevertheless, the outcomes of a phenomenological study broaden the researcher's horizons as it inspires to see ahead and define one's own attitudes, views and beliefs through intentional study of the lived experiences. Although at first glance phenomenology may seem to be somewhat philosophically overwhelming and at the same time complex in its methodical procedures, it provided me with a unique tool to enter the worlds which normally remain uncovered to the 'external' observer. It is time-consuming, can be emotionally draining and cognitively confusing, but the final result, i.e. discovering the others' cultural worlds and the daily lives of their representatives as heard from them and shown by them on the ground of mutual trust and openness, is a truly rewarding experience.

Conclusion

Constructivism, which I used for the guiding paradigm of my research project, helped me retain an open, flexible and dialogic relationship with the object as well as the subjects of my research. Because in constructivist view knowledge is a form of cognitive representation, it is not important 'who possesses the knowledge', as its acquisition is the key matter. For that reason, the activity of the 'learning' researcher leads to constructing and reconstructing the research reality. Therefore, as researchers we are active, subjective and involved in the research process not only cognitively, but also emotionally. The constructivist paradigm gives priority to the activity of the researcher that immerses his/herself in the world of the research, stressing the fact that we tend to perceive the reality through the prism of one's own culture and experiences, hence the researched reality is never objective and free from assigned meanings and contexts. It gives the researcher a great sense of empowerment and meaningfulness. Moreover, it makes the research a truly interactive and dialectic journey, as by entering interaction with others we exchange the visions of the realities and share the diversities of these views, reaching to consensus. I would recommend this approach to researchers who are self-conscious, driven by cultural values and open to non-conventional ways of perceiving the reality around, not within the nomothetic but rather through idiographic, emic perspectives.

The phenomenological research method, as described by Moustakas (1994), that corresponds greatly in its ontological and axiological assumptions with the constructivist approach, allows researchers to enter the hidden, at times intimate, worlds of the researched in a mindful and subjective manner searching for the essence of experiences as well as their meaning, rather than carrying out research by means of various measurements and seeking their explanations. I would recommend phenomenology as a useful qualitative basis to operate in the world of the researched recognising their subjectivity, autonomy and self-reflectivity. Nonetheless, it requires the researcher to tread reflexively and mindfully; hence self-awareness comes across as a key soft skill in this regard. It is an approach directed towards the researched and their world visions, empowering their status as the co-producers of knowledge, demanding that the researcher be a 'humble background' for the evolving narratives s/he stimulates and encourages through dialogue. Constructivism and phenomenology are suitable in all the research designs where we aim at reflexive, subjective and interpretive (re)construction of the social and cultural worlds around us. It might be emotionally and culturally challenging at times, but simultaneously it provides the researcher with a great sense of fulfilment and accomplishment.

References

Ablewicz, K. (1994) *Hermeneutyczno-fenomenologiczna perspektywa badań w pedagogice* [in Polish]. Kraków: Wydawnictwo UJ.

Berger, P. and Luckmann, T. (1967) *The Social Construction of Reality: A Treatise in the Sociology of Knowledge*. Garden City, NY: Doubleday.

Bloor, D. (1996) Idealism and the sociology of knowledge. *Social Studies of Science* 26, 839–856.

Bogdan, R.C. and S.K Biklen (1992) *Qualitative Research for Education: An Introduction to Theory and Methods*. Boston: Allyn & Bacon.

Cobb-Stevens, R. (2018) Hobbes and Husserl on reason and its limits. In R. Sokolowski (ed.) *Edmund Husserl and the Phenomenological Tradition* (pp. 47–62) Washington D.C: The Catholic University of America Press.

Cohen, L., Manion, L. and Morrison, K. (2007) *Research Methods in Education*. New York, NY: Routledge.

Creswell, J.W. (2003) *Research Design: Qualitative, Quantitative and Mixed Methods Approaches* (2nd edn). Thousand Oaks, CA: Sage.

Creswell, J.W. (2007) *Qualitative Inquiry & Research Design Choosing Among Five Approaches*. Thousand Oaks, CA: Sage.

Creswell, J.W. (2012) *Educational Research: Planning, Conducting and Evaluating Quantitative and Qualitative Research* (4th edn). Upper Saddle River, NJ: Pearson.

Creswell, J.W. and Poth, C.N. (2017) *Qualitative Inquiry and Research Design: Choosing among Five Approaches* (4th edn). Thousand Oaks, CA: Sage.

Crotty, M. (1998) *The Foundations of Social Research: Meaning and Perspective in the Research Process*. London: SAGE Publications Ltd.

Devitt, M. (1997) *Realism and Truth*. Princeton: Princeton University Press.

Fochtman, D. (2008) Phenomenology in paediatric cancer nursing research. *Journal of Paediatric Oncology Nursing* 25 (4), 185–192.

Gadamer, H.G. (1997) *Truth and Method*. (2nd rev. ed.) (J. Weinsheimer and D.G. Marshall, Trans. rev.). New York, NY: Continuum. (Original work published 1960).
Gardner, S. (1999) *Kant and the Critique of Pure Reason*. London and New York: Routledge.
Gearing, R. (2004) Bracketing in research: A typology. *Qualitative Health Research* 14 (10), 1429–1452.
Giorgi, A. and Giorgi. B. (2003) Phenomenology. In J.L. Smith (ed.) *Qualitative Psychology – A Practical Guide to Research Methods* (pp. 25–51). London: Sage.
Hacking, I. (1999) *The Social Construction of What?* Cambridge, MA: Harvard University Press.
Heidegger, M. (1996) *Being and Time*. Re-translated by Joan Stambaugh. Albany: State University of York Press.
Honebein, P.C. (1996) Seven goals for the design of constructivist learning environments. In B.G. Wilson (ed.). *Constructivist Learning Environments: Case Studies in Instructional Design* (pp. 11–24). Englewood Cliffs: Educational Technology Publications.
Husserl, E. (1970) *Logical Investigations*. New York: Humanities Press.
Husserl, E. (1977) *Phenomenological Psychology: Lectures, Summer Semester 1925* (J. Scanlon, Trans.). The Hague, Netherlands: Martinus Njihoff. (Original work published posthumously 1962).
Kivunja, C. and Kuyini, A.B. (2017) Understanding and applying research paradigms in educational contexts. *International Journal of Higher Education* 6 (5), 26–41.
Landgridge, D. (2008) Phenomenology and critical social psychology: Directions and debates in theory and research. *Social and Personality Psychology Compass* 2/3 (2008), 1126–1142.
Law, J. and Lodge, P. (1984) *Science for Social Scientists*. London: Macmillan.
Lester, S. (1999) *An Introduction to Phenomenological Research*. Taunton: Stan Lester Developments.Accessed July 27 2018: www.sld.demon.co.uk/resmethy.pdf
Lichtman, M. (2011) *Understanding and Evaluating Qualitative Research*. Thousand Oaks, CA: Sage.
Lincoln, Y. (1990) The making of a constructivist. In E.G. Guba (ed.) *The Paradigm Dialog*. Newbury Park CA: Sage.
Lincoln, Y.S. and Guba, E.G. (2000) Paradigmatic controversies, contradictions, and emerging confluences. In Y.S. Lincoln and E.G. Guba (eds) *Handbook of Qualitative Research* (pp. 163–188). Thousand Oaks, CA: Sage.
Mackenzie, N. and Knipe S. (2006) Research dilemmas: Paradigms, methods and methodology. *Issues In Educational Research* 16, 193–205.
Manning, K. (1997) Authenticity in constructivist inquiry: Methodological considerations without prescription. *Qualitative Inquiry* 3 (1), 93–115.
Merriam, S.B. and Tisdell, E. (2016) *Qualitative Research: A Guide to Design and Implementation*. San Francisco, CA: Jossey-Bass.
Mertens, D.M. (1998) *Research Methods in Education and Psychology: Integrating Diversity with Quantitative Approaches*. London: Sage Publications.
Moustakas, C. (1994) *Phenomenological Research Methods*. London: Sage.
Neumann, W.L. (2000) *Social Research Methods: Qualitative and Quantitative Approaches* (4th edn). Boston: Allyn & Bacon.
Nieswiadomy, R.M. (1993) *Foundations of Nursing Research* (2nd edn). Norwalk, CT: Appleton & Lange.
Qutoshi, S.P. (2018) Phenomenology: A philosophy and method of inquiry. *Journal of Education and Educational Development* 5 (1) June 2018, 215–222.
Rockmore, T. (2008) *On Constructivist Epistemology*. New York: Rowman & Littlefield Publishers

Riemen, D.J. (1986) The essential structure of a caring interaction: Doing phenomenology. In P.M. Munhall and C.J. Oiler (eds) *Nursing Research: A Qualitative Perspective* (pp. 85–105). Norwalk, CN: Appleton-Century-Crofts.

Schwandt, T.A. (1994) Constructivist, interpretivist approaches to human inquiry. In N.K. Denzin and Y.S. Lincoln (eds) *Handbook of Qualitative Research* (pp. 118–137). Thousand Oaks, CA: Sage.

Schwandt, T.A. (2007) *Dictionary of Qualitative Inquiry* (3rd edn). Thousand Oaks, CA: Sage.

Smith D.W. (1993) Transcendental "I". In F.M. Kirkland and D.P. Chattopadhyaya (eds) *Phenomenology: East and West. Contributions to Phenomenology* (In Cooperation with the Center for Advanced Research in Phenomenology), 13. Dordrecht: Springer.

Spiegelberg, H. (1969) *The Phenomenological Movement* (2nd edn). The Hague: Martinus Nijhoff.

Strauss, A. and Corbin, J. (1998) *Basics of Qualitative Research: Techniques and Procedures for Developing Grounded Theory* (2nd edn). Thousand Oaks, CA: Sage.

van Manen, M. (1990) *Researching Lived Experience: Human Science for an Action Sensitive Pedagogy*. London, Ontario: The University of Western Ontario.

Wojnar, D. and Swanson, K. (2007) Phenomenology: An exploration. *Journal of Holistic Nursing* 25 (3), 172–180.

7 Applying the Interpretive Social Science Paradigm to Research on Tourism Education and Training

Yohei Okamoto

> What I knew and who I was at the start of the journey was different from what I knew and who I was at the end.
> (Mackenzie & Ling, 2009: 51)

Introduction

This chapter presents reflections upon my journey towards understanding, developing and practising independent research skills of my own and coming to understand the interpretive social science paradigm. It relates to my Master of Philosophy (MPhil), which focused on a research project that was undertaken between 2014 and 2016. My MPhil thesis was titled: *The Raison d'etre of Tourism Education* (Okamoto, 2016). The study involved a content analysis of tourism descriptors in order to identify philosophical influences underpinning tourism programmes in Australia. It sought to explore the ontology, ideology and epistemology behind those tourism educational programmes. The aim of this chapter is not about presenting the findings of the research; it rather tells a story on the process through which I came to realise that those assumptions relating to philosophical influences are fundamental components of research practice and how I actively considered my philosophical position in order to determine my research methodology so as to align my research approach with a research paradigm. I argue that three philosophical lenses which I used as a framework in my research – *ontological*, *epistemological* and *ideological* – became the essential 'standpoints' that guided and formulated my approach to my research. Figure 7.1 visually illustrates this proposition: a *philosophically minded approach* to tourism research.

Broadly, the chapter first presents a reflective case study on my journey as a tourist, tourism professional, tourism student and tourism academic

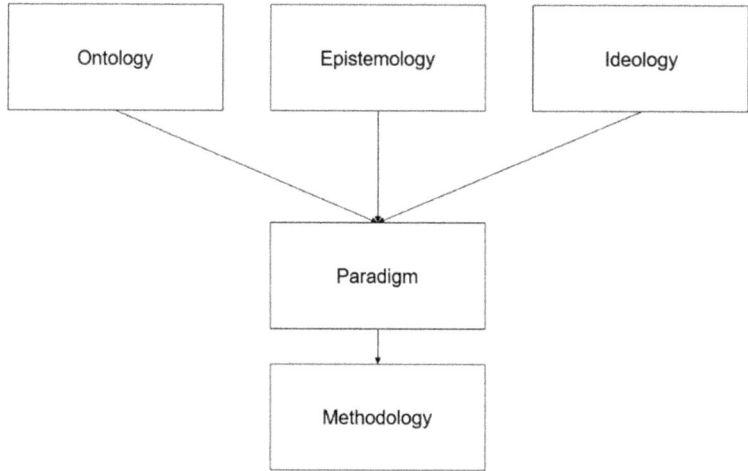

Figure 7.1 A philosophically minded approach to tourism research

and how it has impacted the three philosophical standpoints to be who I am today as an early career scholar. The second part introduces a background to understanding research paradigms and their role in tourism research. The final part explores the interpretivist paradigm and its application to my MPhil investigation in tourism education and training.

My Story

My career and interest in tourism began in New Zealand where I decided to enrol in a vocational course in international tourism as an exchange student from Japan. I was born, grew up and was educated in Japan where tourism was not a common field of study, and I vividly remember the first day of the course, not knowing what to expect to learn. The lecturer told the class that 'tourism is a field of study dedicated to training professionals who can stage a memorable trip of a lifetime'. I immediately fell in love with the description and cemented my future professional pursuit in this field. This was when the ontological notion of the 'tourism industry' was planted in my mind.

I also experienced my first tourism-related work as a customer liaison officer at a tourist information centre. This led to my realisation of the existence of the neoliberal/business ideological nature of tourism, which aligned strongly with my undergraduate study: a Bachelor of Business Administration (BBA). My epistemological approach to tourism knowledge was largely based on the business discipline as well as extra-disciplinary knowledge as my learning was based on business subjects and industry sources. This was also the time when I began travelling extensively around the world, starting with New Zealand, and then going to

India, the USA, Europe and South East Asia. No matter where I visit, the act of travelling has been and still provides an eye-opening or peak experience (Quan & Wang, 2004) which lets me learn about different ways of life, embrace and appreciate diverse natural wonders, and immerse and reflect myself in the intriguing feeling of 'otherness' as a cultural and ethnic minority (Feighery, 2006; Kahn, 2011).

Upon the completion of my undergraduate degree in Japan, I moved to Australia, where I subsequently completed another vocational qualification in tourism: a Diploma in Tourism (DipTou). I then completed two higher education qualifications: a Postgraduate Diploma in Tourism (PostgradDipTou) and a Master of Philosophy (MPhil) at Murdoch University in Perth. During and between my academic journeys in Australia, I worked in various sectors within the tourism industry. Specifically, I worked in the tourism and hospitality sectors as a multilingual tour leader, a travel agent, a hotel receptionist and a sommelier to a tourism research consultant. I began my academic career as a vocational education and training (VET) lecturer in tourism, events, and hospitality after completing my PostgradDipTou and having worked in tourism for over five years in Australia.

My business-oriented epistemological understanding of tourism, coupled with a strong neoliberal ideological influence resulting from my BBA studies and working extensively in the tourism industry, was critically challenged during the academic pursuits at Murdoch University. This was partly because the tourism programme was housed in the School of Arts, where tourism was predominantly seen from a social sciences perspective with a strong focus on sustainability (Murdoch University, 2018). I encountered and was immersed in concepts such as tourism impacts (McKercher, 1993); tourism platforms (Jafari, 1990, 2005); sustainable tourism (Weaver, 2006); tourism ethics (Macbeth, 2005); and the philosophic practitioner (Dredge *et al.*, 2012; Tribe, 2002). More recently, I became interested in tourism and spirituality (Barkathunnisha *et al.*, 2018). All of these perspectives have influenced my understanding of tourism ontologically, epistemologically and ideologically.

The broad theme of my MPhil thesis, post-secondary tourism education and training, emerged in my mind when I began working as a lecturer at a vocational college and had experienced the 'practical world' of tourism in the various sectors within the tourism industry. In addition to this, being a student of tourism also led me to the realisation of the existence of expanding tourism education. There was also a fundamental interest in certain simple yet deep questions such as: *What makes good tourism education and training? Why are there VET and higher education (HE) sectors in tourism education and training? What knowledge and skills should be taught as part of a tourism course?* and *What is the raison d'etre of tourism education?*

Having experienced the dichotomy between the 'practical' and 'theoretical' worlds of tourism (Tribe, 2002), the inquiry into the philosophical aspects of tourism education began with my realisation of the diverse perspectives held by tourism professionals and education practitioners. Such a discussion led to the cross-disciplinary nature of my research project (i.e. tourism and education) and determination of my research paradigm and research positionality as an interpretivist social scientist. Further, given the socially constructed nature of the topic, this study required a critical approach. The term 'critical' was interpreted as being deeply conscious about the socially constructed nature of knowledge; the choices of knowledge people make; and the various philosophical underlining influences. Therefore, by being critical, I acknowledged a notion that it is about recognising and questioning the status-quo and the dominant and ruling ideology. I also found that it gives equal attention to different ideological views regardless of the power and influence of each of the views. My MPhil thesis hence sought to explore the differences and similarities in relation to those three aspects of philosophical underpinnings among and between the VET and HE tourism programmes in Australia. To pursue my MPhil topic, I had to myself understand the terms 'ontology', 'epistemology' and 'ideology'. I here present an overview of these with insights into how they shaped the approach to my research. I will follow this section with a review of the interpretive social science paradigm and how I applied this paradigm to my research.

Ontology

Firstly, ontology can simply be defined as 'the nature of reality' (Jennings, 2010: 36). In a research context, ontology is used in determining the researcher's gaze (Hollinshead, 1999). For instance, a classic illustration of an ontological model of tourism is provided by Leiper (1979) depicting five distinctive elements of a *tourism system*: tourist generating regions, tourist destination regions, transit routes, departing tourists and returning tourists. Here, the term tourism 'system' is deliberately used to differentiate it from the tourism 'industry'. The tourism system relates to the holistic ontological view of tourism as a phenomenon that incorporates the five elements described above and their surrounding environments, while the tourism 'industry' only considers the business activities in relation to tourism. Further, one could ontologically see tourism as a 'workplace' where tourists and tourism workers interact (Tribe, 1999).

Hence, I realised that ontological standpoints or assumptions of tourism can be expressed on a continuum. At one end, tourism can be seen as a narrow view of the workplace, departments and organisations which incorporate the perspectives of workers and operations managers. At the other end, tourism may incorporate the wider perspectives of various stakeholder groups such as government agencies, private businesses, non-profit organisations and communities. Further, I was aware that there are more holistic views of the social and natural environment as well as philosophical and ideological spaces (Higgins-Desbiolles, 2006). This outlook enabled me to understand that tourism is seen as a phenomenon that could be an agent for a better world, self-reflection and development. From this thinking, I concluded that traditional academic fields of research are waiting to be explored and discovered.

Hence, I realised that one's ontological standpoint can be identified by giving careful consideration to the following question: *what is tourism, and what is not tourism, in the context of a research project?* as one (whether a tourism professional, a tourism researcher, a tourism student or even a tourist) can have different understandings of tourism.

Epistemology

Epistemology refers to the theory of knowledge or knowing. It relates to thoughts given to determine the process of knowledge creation, acquisition, utilisation, communication and re-utilisation (or refinement) (King, 2009). Being aware of various epistemological approaches in research and knowledge is important because 'knowledge is never purely objective, nor value or interest-free but rather is subject to a range of sociological forces' (Tribe & Liburd, 2016: 54). Hence, it was critical for me to recognise and reflect on epistemological processes to conceptualise how knowledge relevant to my research was produced, utilised and perceived.

Tribe (1997) provided four epistemological processes in tourism studies: disciplinary (and multidisciplinary), interdisciplinary, business interdisciplinary and extra-disciplinary. The first approach is *disciplinary* knowledge. This is where tourism knowledge can be constructed through a variety of established disciplines such as sociology, social psychology, geography, anthropology and economics (Echtner & Jamal, 1997). *Multidisciplinary* knowledge has greatly contributed to the body of knowledge from the inception of tourism as an academic field of study (Jafari, 1990). However, it also involves limitations as it creates boundaries between knowledge from different disciplines which hinders the cohesive progression of knowledge and creates an un-unified set of knowledge about tourism. This takes tourism away from being a paradigmatic area of study (Echtner & Jamal, 1997).

The second epistemological process is through the *interdisciplinary* approach where knowledge is created through a collaboration between disciplines. Darbellay and Stock (2012: 453) define the interdisciplinary approach as 'a process of mobilising different institutionalised disciplines through a dynamic interaction in order to describe, analyse and understand tourism's complexity'. The critical difference between multidisciplinary and interdisciplinary knowledge is that while multidisciplinary knowledge is created with a single (or single dominant) discipline, the process of interdisciplinary knowledge creation involves interactions, dialogue, integration, synthesis and/or exchange within and across disciplines resulting in the generation of 'new knowledge' that does not belong to a particular discipline but between disciplines (Shearer, 2007).

Third, *business interdisciplinary* knowledge in the context of tourism is knowledge created between business disciplines (e.g. marketing, accounting and economics) and the world of tourism. This type of knowledge has developed due to the ever-increasing economic impact of tourism and tourism scholarship's close affiliation with business disciplines (Belhassen & Caton, 2011). This form of knowledge is both theoretical and applied in nature and has a specific purpose to serve such as economic performativity (Tribe, 2002) or practical problem-solving (Belhassen & Caton, 2009) where the fundamental objective of gaining knowledge is for improved profit, better service and best business practice.

Finally, *extra-disciplinary* tourism knowledge is created without a discipline (Tribe & Liburd, 2016). It is commonly applied, operational and/or highly contextualised in nature (Volgger & Pechlaner, 2014). The knowledge that is generated by the industry is not found in academic journals or validated through rigorous peer review assessment (Tribe, 1997). This knowledge can include tourism jargon (or technical terms), the structure of the tourism industry, the operation of sector-specific software (e.g. Galileo, Sabre and Opera), the use of particular equipment and findings from industry representative bodies and independent research firms. Extra-disciplinary knowledge can also be created by members of the community (Dredge & Schott, 2013). Extra-disciplinary knowledge is often created by talented and altruistically motivated members of the wider public frequently with the aid of the internet, Web 2.0 and Information and Communication Technologies (ICTs) through platforms such as Wikipedia, the Conversation, Quora, TED and social media, allowing highly social and collaborative knowledge creation (Liburd & Christensen, 2013; Prasarnphanich & Wagner, 2009). The emergence of such self-organising and open-access knowledge creation is relatively new in tourism research; however, it is not within the bounds of the traditional discipline. Yet this type of knowledge creation is growing and is influencing the face of tourism research (Coles *et al.*, 2006).

Hence, I believe that the main rationale for the consideration of one's epistemological standpoint is to establish *what type(s) or form(s) of knowledge you consider 'credible' for the context of your research.* Giving careful thought to one's epistemological approach informs the types of knowledge required for the research. A tourism researcher can set his or her epistemological standpoint to determine what type(s) of knowledge should be considered 'vigorous' and, therefore, incorporated in his or her research project.

Ideology

Ideology can be defined as a 'system of thoughts that are generally internally consistent and pertain to an understanding of the interrelationship between state, market and society' (Webster & Ivanov, 2016: 109). According to Collier (1982: 13), 'an ideology serves the interests of some part of society ... from which it is argued that the state of society ought to be changed or preserved in certain respects'. It is also used to describe specific coherent subsets of beliefs commonly referred to as '–isms' (Tribe, 2006). Examples are conservatism, liberalism, anarchism, Marxism, fascism and environmentalism, to name a few (Webster & Ivanov, 2016). Ideologies are the forces that affect one's intentions, values, cultures and powers that exist at the macro (regional, national and international), meso (institution-wide) and micro (with peers/family members) levels (Barnett, 2013; Dredge *et al.*, 2013). Micro-level systems can be your values, personalities and (taken-for-granted) beliefs associated with your research (Boyle, 2015). This level of ideology can overlap with 'axiology' which is the study of values and ethics (Edelheim, 2014). This aspect of tourism research has been examined by Tribe's (2010) paper on 'academic tribes and territories' in tourism academia. He argues that the current state of tourism as a field of study has been, and still is, an ideologically fragmented academic field due to the existence of 'tourism academics' with different backgrounds; each bringing their own rules, hierarchies and cultures to their research. His research highlights 'the tensions between industries vs. blue skies research, individuals vs. institutions, new guard vs. old guard, business vs. social science and those who feel outsiders vs. insiders' (Tribe, 2010: 31).

Further, McKercher and Prideaux (2014: 25) argue that tourism research is responsible for the distribution of various types of 'academic myths' often led by certain ideologies. They argue that while these myths can be beneficial in supporting 'the raison d'etre for a field of study and for providing a central rallying point for researchers', many of them are based on 'wishful thinking and cannot be based on rigorous evaluation' (McKercher & Prideaux, 2014: 27).

Therefore, I came to the realisation that ideologies have a strong influence on one's research approach as the ideological thoughts that include intentions, values, cultures and powers can play a significant role in the shaping of the one's research. Hence, ideological questions one may ask include: *What tourism do you want to see? Who (or what) is driving you to conduct the research?* or even *Why do you do this research?* Ideological influences can, therefore, be forces that underpin the directions of research that exist in a multitude of ways in society. For me, the three components of ontology, epistemology and ideology, are what I consider as the three fundamental cornerstones of tourism research, and are reflected in the development of my researcher personality and social science paradigm.

Review of the Interpretive Social Science Paradigm

In any research project, it is vital to consider the possible theoretical research paradigms as they inform the methodological approach of the investigation and how these can result in entirely different ways of conducting the investigation (Tribe, 2001). In other words, research paradigms can be understood as the 'standpoint' of a researcher. It is also important to note that a research paradigm is a 'mindset' or 'viewpoint' that guides the researcher; it does not, or should not, restrict how research is conducted as limiting methods can 'potentially diminish and unnecessarily limit the depth and richness of a research project' (Mackenzie & Knipe, 2006: 200). Paradigms are something that a researcher should consider at an earlier stage of the research process. Hence, they are generally discussed in early chapters of research texts (for example, Babbie, 2011; Jennings, 2010; Walter, 2013). This poses a challenge for many PhD candidates and early career researchers (ECRs) as the impacts of paradigms on research projects are not often fully appreciated at the beginning of research (Mackenzie & Ling, 2009). In tourism studies, the emergence of the tourism field led to the 'knowledge-based' platform and to predominantly positivist approaches up until the early 2000s (Jafari, 2001: 31). Since then there has been a shift in the wider acceptance and appreciation of other alternative paradigms such as the interpretivist social science paradigm and critical theory, which focus more on a qualitative and value-laden approach (Caton, 2012; Macbeth, 2005). Now, there are a number of commonly employed paradigms in tourism research which include positivism, postpositivism, interpretivism, critical theory, participatory theory and feminist theory, to name a few (Jennings, 2010).

The interpretive social science paradigm has its origin in Max Weber's 'verstehen' or empathetic understanding from the late 19th century (Jennings, 2010: 40) as well as phenomenology and hermeneutics (Mackenzie & Knipe, 2006). It encompasses research that is exploratory and qualitative and identifies multiple underlying meanings and realities rather than looking for a 'true' single reality or answer that positivist scientists would seek (Jennings, 2010). The interpretivist paradigm surged in the social sciences in the postmodern era in the 1980s as an alternative to the dominant scientific and objective philosophical perspectives of positivism and post-positivism of the time (Dean, 2018; Wall, 2006). According to Tribe (2001), the paradigm is suitable for research methods such as observation, case studies and literary criticism. Further, this paradigm is generally associated with the inductive approach where no hypothesis is formed prior to data collection. Accordingly, the explorative nature of the paradigm, is seen as advantageous to make new discoveries and theoretical insights (Altinay et al., 2015).

Reflections on Applying the Interpretive Social Science Paradigm

The focus of this chapter now turns back again to my journey, this time as a researcher. This reflection describes the process involved in aligning my research methodology with a research paradigm to develop my researcher personality.

My MPhil investigation involved a content analysis of the descriptors of 291 HE and VET units of study (161 and 130, respectively) in Australia. Content analysis involves 'careful, systematic and detailed examination of material in an effort to identify patterns, themes, biases and meanings' (Camprubí & Coromina, 2016: 134). Further, considering the nature of the enquiry, the *interpretive social science paradigm* was aligned to my study. My research was undertaken in an inductive manner where no hypothesis was formed; instead, I began with observations, followed by finding a pattern of the phenomenon, and finally, a conclusion was drawn (Babbie, 2011). Given that content analysis is a flexible methodology which can be performed in both a quantitative and a qualitative manner (Bengtsson, 2016), three types of coding, as suggested by Richard (2009) were adopted, namely *descriptive coding* which involved the identification of easily identifiable attributes, *topic coding* to identify the content and *analytical coding* to identify influences and meanings.

Further, considering my positionality as a researcher, I took an 'emic' perspective where the researcher is an insider and subsequently experiences the phenomena of the investigation; taking on the beliefs, attitudes and other points of view shared by the 'real' members (Babbie, 2011: 320). In social anthropology, this view is also called an 'artistic' view as opposed to a 'scientific' view as the emic perspective can better appreciate culture and language as well as assist in explaining humanistic values such as life, motives, interests, conflict and the personality of specific actors (Walle, 1997). This view was taken due to my profession as a VET teacher as well as a postgraduate student, allowing me to see both sides of the phenomenon as a provider (teacher) and receiver (student) of tourism education.

I consider that my alignment with the interpretivist paradigm and content analysis for my research approach was suitable for my project. I have previously experienced tourism programmes in the VET and HE sectors (DipTou and PostgradDipTou) as both a student and an educator and feel that I could provide valuable interpretation of any data collected. I am one of the first cohort of 'Generation T'; that is, people whose education has involved tourism, tourists and related topics and generally have multiple points of interests including sciences and business (Pearce, 2011). Further, I have worked in the tourism industry, and I was already a teaching practitioner in the HE sector in Australia prior to my postgraduate studies. These experiences enabled me to have an emic (insider) perspective producing subjective, yet critical, observations on the current provision of tourism education programmes. Further, performing a content analysis with multiple levels of coding allowed a comprehensive immersion into the research and development of understanding of the tourism units that extended from the surface to deeper and more meaningful complexities (Bengtsson, 2016).

Conversely, I experienced multiple issues in relation to aligning the interpretivist paradigm with the research approach. First, although I aligned my research with the interpretive social science paradigm, this method can be highly subjective and come with a cost to reliability (Babbie, 2011). In an attempt to balance subjectivity, quantitative data were also integrated into the presentation of the results. That way, the research could provide both structures; subjective reflection and interpretation of data. In addition, there is the issue of researchers' bias which can be appreciated as 'clouding of interpretation' by interpretivist researchers (Boyle, 2015: 64). When content analysis is used for the interpretation of qualitative data, the findings are subject to researchers' biases, subjectivity and lack of generalisability (Dean, 2018), and my investigation was no exception to this. In order to minimise the unwanted subjectivity bias, the instruments were tested and verified by my supervisors (Camprubí & Coromina, 2016). Finally, disclosing my positionality allowed me to sufficiently reflect my rather unique gaze as a Japanese, Generation T, early career researcher and educator on the topic of the raison d'etre of tourism education. Given the socially constructed nature of the tourism education curriculum space (Dredge *et al.*, 2012; Tribe, 2014), I believe this conscious practice of 'entangling' my lived experience and philosophical standpoints in the research let to the creation of reflexive knowledge and the expression of critical voice from outside the dominant discourses (Ateljevic *et al.*, 2005).

Conclusion

This chapter has presented my proposition that three philosophical lenses (ontological, epistemological and ideological) have a profound impact on one's choice of a research paradigm and methodology. I have used my own reflective personal case study to outline my research journey where I aligned with the interpretive social science paradigm.

As we all live and work in diverse, contested, fragmented, and ideologically entwined worlds, everyone has a unique life story that forms part of who they are as a person and as a researcher. I suggest that using the philosophically minded approach I have presented here will allow you to analyse yourself through these philosophical lenses and may allow you to identify influences that shape who you are and who you wish to be as a researcher.

References

Altinay, L., Paraskevas, A. and Jang, S. (2015) *Planning Research in Hospitality and Tourism* (2nd edn). Routledge.

Ateljevic, I., Harris, C., Wilson, E. and Collins, F.L. (2005) Getting 'entangled': Reflexivity and the 'critical turn' in tourism studies. *Tourism Recreation Research* 30 (2), 9–21. https://doi.org/10.1080/02508281.2005.11081469

Babbie, E. (2011) *The Basics of Social Research* (5th edn). Wadsworth: Cengage Learning.

Barkathunnisha, A.B., Lee, D., Price, A. and Wilson, E. (2018) Towards a spirituality-based platform in tourism higher education. *Current Issues in Tourism* 22 (17), 2140–2156. https://doi.org/10.1080/13683500.2018.1424810

Barnett, R. (2013) *Imagining the University*. Abingdon: Routledge.

Belhassen, Y. and Caton, K. (2009) Advancing understandings: A linguistic approach to tourism epistemology. *Annals of Tourism Research* 36 (2), 335–352. https://doi.org/10.1016/j.annals.2009.01.006

Belhassen, Y. and Caton, K. (2011) On the need for critical pedagogy in tourism education. *Tourism Management* 32 (6), 1389–1396. https://doi.org/10.1016/j.tourman.2011.01.014

Bengtsson, M. (2016) How to plan and perform a qualitative study using content analysis. *NursingPlus Open* 2, 8–14. https://doi.org/10.1016/j.npls.2016.01.001

Boyle, A.R. (2015) Space for sustainability?: From curriculum to critical thinking in Australian tourism higher education thinking in Australian tourism higher education [PhD Thesis, Southern Cross Unversity]. https://epubs.scu.edu.au/theses/444/

Camprubí, R. and Coromina, L. (2016) Content analysis in tourism research. *Tourism Management Perspectives* 18, 134–140. https://doi.org/10.1016/j.tmp.2016.03.002

Caton, K. (2012) Taking the moral turn in tourism studies. *Annals of Tourism Research* 39 (4), 1906–1928. https://doi.org/10.1016/j.annals.2012.05.021

Coles, T., Hall, C.M. and Duval, D.T. (2006) Tourism and post-disciplinary enquiry. *Current Issues in Tourism* 9 (4), 293–319. https://doi.org/10.2167/cit327.0

Collier, K.G. (1982) Ideological influences in higher education. *Studies in Higher Education* 7 (1), 13–19. https://doi.org/10.1080/03075078212331379271

Darbellay, F. and Stock, M. (2012). Tourism as complex interdisciplinary research object. *Annals of Tourism Research* 39 (1), 441–458. https://doi.org/10.1016/j.annals.2011.07.002

Dean, B.A. (2018) The interpretivist and the learner. *International Journal of Doctoral Studies* 13, 1–8. https://doi.org/https://doi.org/10.28945/3936

Dredge, D., Benckendorff, P., Day, M., Gross, M.J., Walo, M., Weeks, P. and Whitelaw, P. (2012) The philosophic practitioner and the curriculum space. *Annals of Tourism Research* 39 (4), 2154–2176. https://doi.org/10.1016/j.annals.2012.07.017

Dredge, D., Benckendorff, P., Day, M., Gross, M.J., Walo, M., Weeks, P. and Whitelaw, P. (2013) Drivers of change in tourism, hospitality, and event management education: An Australian perspective. *Journal of Hospitality and Tourism Education* 25 (2), 89–102. https://doi.org/10.1080/10963758.2013.805091

Dredge, D. and Schott, C. (2013) Academic agency and leadership in tourism higher education. *Journal of Teaching in Travel and Tourism* 13 (2), 105–129. https://doi.org/10.1080/15313220.2013.786312

Echtner, C.M. and Jamal, T.B. (1997) The disciplinary dilemma of tourism studies. *Annals of Tourism Research* 24 (4), 868–883. https://doi.org/10.1016/S0160-7383(97)00060-1

Edelheim, J.R. (2014) Ontological, epistemological and axiological issues. In D. Dredge, D. Airey and M.J. Gross (eds) *Routledge Handbook of Tourism and Hospitality Education* (pp. 30–42). Abingdon: Routledge.

Feighery, W. (2006) Reflexivity and tourism research: Telling an (other) story. *Current Issues in Tourism* 9 (3), 269–282. https://doi.org/10.2167/cit/mp006.0

Higgins-Desbiolles, F. (2006) More than an "industry": The forgotten power of tourism as a social force. *Tourism Management* 27 (6), 1192–1208. https://doi.org/10.1016/j.tourman.2005.05.020

Hollinshead, K. (1999) Surveillance of the worlds of tourism: Foucault and the eye-of-power. *Tourism Management* 20 (1), 7–23. https://doi.org/10.1016/S0261-5177(98)00090-9

Jafari, J. (1990) Research and scholarship: The basis of tourism education. *The Journal of Tourism Studies* 1 (1), 33–41.

Jafari, J. (2001) Scientification of tourism. In V.L. Smith and M. Brent (eds) *Hosts and Guests Revisited* (pp. 28–41). New York: Cognizant Communication Corporation.

Jafari, J. (2005) Bridging out, nesting afield: Powering a new platform. *Journal of Tourism Studies* 16 (2), 1–5.

Jennings, G. (2010) *Tourism Research* (2nd edn). Milton: John Wiley & Sons Australia.

Kahn, S. (2011) Putting ethnographic writing in context. *Writing Spaces: Readings on Writing* 2, 175–192. https://wac.colostate.edu/books/writingspaces2/kahn—putting-ethnographic-writing.pdf

King, W.R. (2009) Knowledge management and organizational learning. *Annals of Information Systems* 4, 3–13. https://doi.org/10.1007/978-1-4419-0011-1

Leiper, N. (1979) The framework of tourism: Towards a definition of tourism, tourist, and the tourist industry. *Annals of Tourism Research* 6 (4), 390–407. https://doi.org/10.1016/0160-7383(79)90003-3

Liburd, J.J. and Christensen, I.M.F. (2013) Using web 2.0 in higher tourism education. *Journal of Hospitality, Leisure, Sport and Tourism Education* 12 (1), 99–108. https://doi.org/10.1016/j.jhlste.2012.09.002

Macbeth, J. (2005) Towards an ethics platform for tourism. *Annals of Tourism Research* 32 (4), 962–984. https://doi.org/10.1016/j.annals.2004.11.005

Mackenzie, N. and Knipe, S. (2006) Research dilemmas: Paradigms, methods, and methodologies. *Issues in Educational Research* 16 (2), 193–205.

Mackenzie, N. and Ling, L.M. (2009) The research journey: A Lonely Planet approach. *Issues in Educational Research* 19 (1), 48–60.

McKercher, B. (1993) Some fundamental truths about tourism: Understanding tourism's social and environmental impacts. *Journal of Sustainable Tourism* 1 (1), 6–16. https://doi.org/10.1080/09669589309514793

McKercher, B. and Prideaux, B. (2014) Academic myths of tourism. *Annals of Tourism Research* 46, 16–28. https://doi.org/10.1016/j.annals.2014.02.003

Murdoch University. (2018) *Tourism and Events: Bachelor of Arts and (BA) in Tourism and Events*. https://www.murdoch.edu.au/study/courses/course-details/Tourism-and-Events-(BA)#

Okamoto, Y. (2016) The raison d'être of tourism education: Ontological, epistemological and ideological comparison between the Higher Education and Vocational Education and Training Tourism Programs in Australia [MPil Thesis, Murdoch University]. http://researchrepository.murdoch.edu.au/id/eprint/35243/

Pearce, P.L. (2011) *The Study of Tourism: Foundations from Psychology*. Bingley: Emerald Group Publishing.

Prasarnphanich, P. and Wagner, C. (2009) The role of wiki technology and altruism in collaborative knowledge creation. *Journal of Computer Information Systems* 49 (4), 33–42. https://doi.org/10.1007/s12130-009-9068-x

Quan, S. and Wang, N. (2004) Towards a structural model of the tourist experience: An illustration from food experiences in tourism. *Tourism Management* 25 (3), 297–305. https://doi.org/10.1016/S0261-5177(03)00130-4

Richard, L. (2009) *Handling Qualitative Data: A Practical Guide* (2nd edn). London: Sage Publications.

Shearer, M.C. (2007) Implementing a new interdisciplinary module: The challenges and the benefits of working across disciplines. *Practice and Evidence of Scholarship of Teaching and Learning in Higher Education* 2 (1), 2–20.

Tribe, J. (1997) The indiscipline of tourism. *Annals of Tourism Research* 24 (3), 638–657.

Tribe, J. (1999) The concept of tourism: Framing a wide tourism world and broad tourism society. *Tourism Recreation Research* 24 (2), 75–81. https://doi.org/10.1080/02508281.1999.11014879

Tribe, J. (2001) Research paradigms and the tourism curriculum. *Journal of Travel Research* 39 (4), 442–448. https://doi.org/10.1177/004728750103900411

Tribe, J. (2002) The philosophic practitioner. *Annals of Tourism Research* 29 (2), 338–357. https://doi.org/10.1016/S0160-7383(01)00038-X

Tribe, J. (2006) The truth about tourism. *Annals of Tourism Research* 33 (2), 360–381. https://doi.org/10.1016/j.annals.2005.11.001

Tribe, J. (2010) Tribes, territories and networks in the tourism academy. *Annals of Tourism Research* 37 (1), 7–33. https://doi.org/10.1016/j.annals.2009.05.001

Tribe, J. (2014) The Curriculum: A philosophic practice. In D. Dredge, D. Airey and M.J. Gross (eds) *Routledge Handbook of Tourism and Hospitality Education* (p. 17). Abingdon: Routledge.

Tribe, J. and Liburd, J.J. (2016) The tourism knowledge system. *Annals of Tourism Research* 57, 44–61. https://doi.org/10.1016/j.annals.2015.11.011

Volgger, M. and Pechlaner, H. (2014) Transdisciplinarity and postdisciplinarity in tourism and hospitality education. In D. Dredge, D. Airey and M.J. Gross (eds) *Routledge Handbook of Tourism and Hospitality Education* (pp. 85–102). Abingdon: Routledge.

Wall, S. (2006) An autoethnography on learning about autoethnography. *International Journal of Qualitative Methods* 5 (2), 146–160. https://doi.org/10.1177/160940690600500205

Walle, A.H. (1997) Quantitative versus qualitative tourism research. *Annals of Tourism Research* 24 (3), 524–536. https://doi.org/10.1016/S0160-7383(96)00055-2

Walter, M. (2013) *Social Research Methods* (3rd edn). Oxford: Oxford University Press.

Weaver, D. (2006) *Sustainable Tourism: Theory and Practice*. Oxford: Elsevier.

Webster, C. and Ivanov, S. (2016) Political ideologies as shapers of future tourism development. *Journal of Tourism Futures* 2 (2), 109–124. https://doi.org/10.1108/JTF-05-2015-0029

8 An Autoethnographic Chronicle on the Ethnographic Exploration of the Nature of Hotel Work and Hospitality in Far North Queensland

Josephine Pryce

Introduction

This chapter chronicles my engagement with the interpretivist approach of ethnography in my PhD journey and discusses how utilisation of triangulation of methods enabled me to develop an appreciation of the value of ethnography. Within the broader context of a phenomenological framework my PhD examined the influence of organisational culture (OC) on the service predispositions (SPs) of hotel workers. The study focused on perspectives of frontline hotel workers in North Australia and aimed to determine organisational factors that contributed to their delivery of hospitality service. The chapter utilises an autoethnographic narrative to reveal how the phenomenological lens of ethnography was considered to be a suitable option for this research.

Post-PhD, my reflections have led to insights that an autoethnographic approach could have equally been applied to my PhD research. Autoethnography is defined as, 'an approach to research and writing that seeks to describe and systematically analyse (graphy) personal experience (auto) in order to understand cultural experience (ethno)' (Ellis et al., 2011: 273). I have thought that perhaps such an autoethnographic approach would have been preferable as it would have allowed for my own (the researcher's) participatory and reflexive autoethnographic stand and voice to come to the fore and so provide more meaningful insights and understanding of participants' attitudes and behaviours and of the nature of hospitality. Nonetheless, the ethnographic approach allowed me to advance knowledge and understanding of hotel work, hospitality and the related work environment. More specifically, it opened my mind to thinking of research beyond what had been, until then, my positivist outlook on research.

This chapter depicts the journey of how I began my PhD with a positivist outlook and progressively warmed to using a triangulation of methods that allowed for ethnographic explorations and ultimately opened up a whole new way of learning about the world of hospitality. In that process my worldview extended and I came to accept interpretivism as a paradigm of relevance, with ethnography as an epistemology that was well suited to my research. In this chapter, I discuss how the interpretivist paradigm was applied in my PhD and present a review of ethnography as a research approach that allows for use of both qualitative and quantitative methods. It details the ontological shift in my thinking and my experience in gaining understanding of 'ways of knowing' and 'ways of doing' that integrated to build knowledge.

The Journey to Find My Paradigm

My memory of when I was undertaking my PhD is that there was not much thought given by research students in my disciplinary area of organisational behaviour to considering one's research paradigm. If there was, it was not thought to be a key part of research design or PhD journeys. The hallowed halls of academia seemed impervious to conversations of paradigms, ontologies and epistemologies. There was, however, an emerging focus in organisational research on interpretivist approaches and triangulation of methods with advice that PhD research should ideally include both quantitative and qualitative approaches. Within the scope of the latter, use of ethnographic approaches in organisational studies was gaining interest among scholars.

Having previously practiced as a natural scientist, I was comfortable and conversant with quantitative research techniques, and as a scholar I was open to exploring and engaging with novel ways to capture and understand the world around us. It was only years later that I realised that when I began my PhD journey, my worldview was contained by the positivist paradigm, wherein impartiality and objectivity are paramount.

Lee's (1991: 342) work was especially pivotal to me shifting my way of thinking. He argued that positivism and interpretivism shared, 'a particular common ground' despite 'the fundamental concepts' that governed each. He maintained that the two approaches are integrated through three levels of understanding: subjective, interpretive and positivist. He built on the work of Alfred Schutz to propose that these levels capture respectively: the observed people and their behaviours (first level); the researcher's interpretation (or reading) of the people being studied (second level); and the researcher's propositions (or hypotheses) relating to the people being studied that are advanced for testing or experimentation (third level). Lee (1991) maintained that 'the methodological legitimacy' of each should be recognised as well as 'the legitimacy of their integration and collaboration'; and he espoused that work of positivist researchers may benefit from using interpretive approaches and vice versa (Lee, 1991: 342–343). Lee (1991: 342) contended that positivism and interpretivism are not necessarily 'opposed and irreconcilable . . . [but are] mutually supportive, rather than mutually exclusive'.

The interpretivist approach in organisational research seeks to study 'social reality' and to understand how 'people create and attach their own meanings to the world around them and to the behaviour that they manifest in that world' (Lee, 1991: 347). What this means is that 'the same physical artifact, the same institution, or the same human action, can have different meanings for different human subjects, as well as for the observing social scientist' (Lee, 1991: 347). The role of researchers is to 'interpret this empirical reality' in relation to what it means to participants and so, 'the social scientist must collect facts and data describing . . . the purely objective, publicly observable aspects of human behaviour . . . [and] the subject meaning this behaviour has for' the people being observed. The result is 'intersubjectively created meanings' which are not captured in the physical reality and methods of the natural sciences, but which demand their own different epistemologies (e.g. Dharamsi & Charles, 2011; Fetterman, 2019; Lee, 1991). Among such epistemologies is ethnography.

At the time when I was undertaking my PhD, it was perceived that ethnography was purely qualitative. In time, it has been acknowledged that it can utilise plurality of methods across qualitative and quantitative domains. Ultimately, all data, whether quantitative or qualitative, can be measured and classified in order to provide general explanatory laws. Even qualitative data can be analysed and interpreted using structured content analyses aided by computer software packages.

The more I read about ethnography, the more I became convinced that it was appropriate for my research. It seemed to lend itself well to exploring the working lives of people in the hotel industry. I was particularly inspired by my own experiences of working in hospitality. At the time of undertaking my PhD, I had already been working in hotels for five years. While completing an Associate Diploma in Business (Hospitality Management), I had gained work in hotels and continued to do so for about a year after I had completed my PhD, a period of over 10 years in total. I enjoyed the 'buzz' of the industry and the work at the frontline where I had opportunity to meet people from all over the world and immerse myself in their adventures as they told stories of their travels. I also revelled in the camaraderie shared by hospitality workers from a range of establishments in town. We were quite an eclectic group of people as we moved from hotel to hotel and knew each other or knew someone who knew someone we knew. Even to this day, many years later, I meet someone at the shopping centre or in town whom I worked with or knew from my time in hospitality and we greet each other like long-lost friends. Hence, it was that I embraced the interpretivist's worldview and that I welcomed the idea of using ethnography.

Application of Interpretivism in My PhD Research

The overall aim of my PhD study was to determine the influence of organisational culture OC on employees' SPs within the context of the hospitality industry and within North Queensland.

In an attempt to examine the influence of the OC on the SPs of hospitality workers, data were gathered from six four-star hotels in North Queensland, Australia. Research into the key components of SPs was emergent at the time and while there was a plethora of research into OC, there were no studies that had investigated the relationship between OC and employees' SPs. First, profiles of SPs were developed. Second, the nature and characteristics of OC were examined. Third, the relationship between SPs and OC was investigated.

A triangulation of research methods was used to achieve these objectives, utilising quantitative and qualitative approaches. The quantitative approach was based on the administration of a questionnaire, which included a version of the Service Predisposition Instrument adapted from the work of Lee-Ross (2000). It included a demographics section and the OC instrument that had been built from a review of the literature. The qualitative approach was ethnographic.

Use of both quantitative and qualitative approaches contributed to a holistic and rich profile of the working lives of hotel workers. The various research approaches triangulated to deliver a cohesive research design. The research strategy was a progressive and iterative process, using multiple sources of information and with each method building upon and providing a more detailed focus than the previous one. This multi-method approach is comparable to mixed methods as described by Creswell and Creswell (2017) but the two should not be confused. At the time of my PhD, I was unaware of mixed methods. Its popularity emerged later.

Justification for the research methods was rooted in ideologies of social sciences where the either of the two paradigms of positivism and interpretivism are considered to be suitable in the study of the constructs of 'service predispositions' and 'organisational culture'. Use of a questionnaire to measure SPs and OC incorporated a positivist orientation, while conducting interviews and use of participant observation made use of an interpretivist approach.

I also found that several authors advocated for the use of multiple methods (i.e. 'triangulation') in research relating to social sciences, such as the study of OC (Hayes, 1997; Rousseau, 1990; Sechrest & Sidani, 1995; Tucker *et al.*, 1990). Patton (1990) contended that the use of triangulation of methods can increase the overall validity of the findings and add depth (from the richness of data obtained) and breadth (allowing generalisation of results obtained from quantitative methods) to the study. Previous studies on hotel work (e.g. Shamir, 1975; Riley & Dodirill, 1992) suggested that the attitudes (i.e. service predispositions) of hotel workers are complex characteristics. Hence, a research design which could extract these complexities was worthy of being used. The triangulation used allowed me to build understanding of the complexities of the social phenomena of SPs and OC.

Ultimately, Schein's (1992) summation of the applicability of qualitative approaches to the study of OC was convincing to me of the value of this approach. He states that cultures are unique, intangible and largely subconscious and that a qualitative approach allows researchers to explore all these aspects of OC. However, in the early days of my PhD and as a novice to qualitative methods, I was still concerned with one of the principal disadvantages of qualitative research in that the analysis and interpretation of the qualitative data is subjective, i.e. made by the researcher (Carlson *et al.*, 2000), and that it can be a lengthy process (Ott, 1989). So, I read on and gained justification for application of qualitative approaches to the study of OC. As a bonus, I found that use of ethnographic approaches was also considered appropriate for assessing the attitudes of hotel workers because it would allow me, the researcher, to understand and explain the complexities of attitudes more fully.

Another dilemma that arose which required deliberate consideration was that with OC, one of the questions raised relates to the appropriate level of analysis: the individual, the workgroup, the department or subunit or the organisation. In explaining the behaviour of people, psychologists use both idiographic (individual) and nomothetic (group) methods (Mullins, 1996). Reeves *et al.* (2013: 1365) define ethnography as 'the study of social interactions, behaviors, and perceptions that occur within groups, teams, organisations, and communities'. This is supported by traditional organisational research that has traditionally utilised a nomothetic approach, which has allowed for discovery of 'generalisable laws or propositions of behavior by averaging responses for each separate variable while using group-comparison designs' (Davis & Luthans, 1984: 243). Researchers who adopt nomothetic research utilise positivist methodologies. While this research has been useful and added to our understanding of organisational social structures and behaviour, it is argued that there is also a need for idiographic research which examines variables and behaviour of interest in their natural setting. One of the early investigations using an idiographic approach was that by Davis and Luthans (1984) as they studied leadership using qualitative methods. Contemporary thought suggests that nomothetic and idiographic approaches can be complementary rather than polarising and belong to both the natural and social sciences (e.g. Korulczyk *et al.*, 2019). The triangulation of methods allowed me to overcome the dilemma of capturing the nature of OC and the characteristics of SPs, with the ethnographic approach offering opportunity to enhance the study of OC and SPs.

An Ethnographic Approach

Ethnography is a phenomenological, qualitative research approach or process that draws on several techniques and at times combines a variety of methods such as observation and interview (Reeves *et al.*, 2013). Its aim is to 'see the world through the eyes of those being researched, allowing them to speak for themselves' (Ticehurst & Veal, 1999: 104). Generally, it involves researchers immersing themselves in a culture, workplace or organisation that may be familiar or foreign to them and becoming part of that group. Bowling (2002: 352) states that ethnography is 'a method of naturalistic enquiry' which focuses on studying people in their natural settings. She adds that it provides 'a descriptive account of social life and culture in a defined social system, based on qualitative methods (e.g. detailed observations, unstructured interviews, analysis of documents)' (Bowling, 2002: 352). This approach allows the researcher to gain a better 'understanding of the (values,) meaning and importance that members of the group impart to their own behaviour and the behaviour of others' (Ticehurst & Veal, 1999: 104).

The word 'ethnography' is derived from the Greek word 'ethnos', which means people, nation or cultural group; and 'graphy', which means writing (Reeves *et al.*, 2013). Traditionally, ethnography is associated with the disciplines of anthropology and sociology where it is used to study culture. Researchers such as Bronislaw Malinowski and Franz Boas were among some of the first anthropology ethnographers of the early 1990s (Reeves *et al.*, 2013). It is thought that the first ethnography was a study on culture and cultural behaviour (such as rituals and practices) that was presented by Malinowski in 1922 (Dharamsi & Charles, 2011). His study was embedded in social anthropology. More recently, it has been acknowledged that ethnography can be used to 'describe and explain a range of social phenomena' within various social groups and 'paint a portrait of the ways' in which these groups 'create meanings' from their daily activities, interactions, experiences, learnings and ways of life that are passed on through generations (Dharamsi & Charles, 2011: 378). So, ethnography is now used in a range of disciplines, including education, psychology and business studies.

Ott (1989) recommends the use of ethnography in OC research. He argues that such an approach enables the researcher to uncover the deeper, more 'invisible level of basic assumptions' that is recognised by scholars (e.g. Hofstede, 1994; Schein, 1985) as being at the invisible core of OC. Dharamsi and Charles (2011: 378) note that ethnography:

> is used to gain a deeper understanding of human behaviour, motivation, and social interaction within specific and complex cultural contexts . . . social complexities . . . ethnographies provide a deeper insight into a

culture, where . . . culture is defined as the collective assumptions and beliefs that influence the practices of a particular group of people who share a social space.

This suggests that the ethnographer is interested in understanding the complexities of people's lives, not just what they say but also what they do, what they think and what they believe and/or value. By participating in the culture, society or organisation, the ethnographer is immersed in trying to understand the people from their perspectives and providing an insider account. Hammersley and Atkinson (1995: 1) note that:

> In its most characteristic form...[ethnography] involves the ethnographer participating, overtly or covertly, in people's daily lives for an extended period of time, watching what happens, listening to what is said, asking questions- in fact, collecting whatever data are available to throw light on the issues that are the focus of the research.

In OC research, studies using ethnographic approaches are limited (Schultz, 1995). Some have used ethnography on its own (e.g. Barley, 1983; Whyte, 1948). Others have used this methodology in combination with other approaches (e.g. Siehl & Martin, 1984). Some have been conducted in the hotel industry. For example, Whyte (1948) used participant observation and interview methods to conduct an ethnographic study of the restaurant industry in Chicago. He examined individuals' attitudes in an attempt to uncover not only the restaurant structure but also the human side, and its problems, of the restaurant sector. It is argued that these ethnographic approaches to studying cultures have several advantages. They can provide depth and richness to a study by tapping into nonverbal cues such as signs and symbols. For example, Barley (1983) used a semiotic approach, to study the significance of signs in funeral homes. He found that when corpses were laid out for viewing, signs were used or created around the corpse so as to give viewers the impression that the person was asleep. The impression was 'to simulate a lifelike appearance' (Barley, 1983: 402). Ethnographic methods, when used in conjunction with other research methods can also provide profound insights into OC. For example, Siehl and Martin (1984) used interviews to supplement findings from questionnaires to uncover the enculturation process of new sales staff and compare such processes between groups within the organisation and across time frames.

For my PhD, by working in the industry, I was afforded the opportunity to take an ethnographic approach that provided an emic (insider's view) of the hotel industry and so offered fuller insights into and understandings of employees' attitudes, OC and the relationship between the two. Working in the industry meant that I was able to observe behaviour and ask questions relating to the SPs and the OC themes that were being explored in my PhD study. Such an approach proved to be a powerful way of enriching this study because in addition to adding depth to the findings that had been drawn out of the quantitative technique (i.e. use of questionnaire), other factors emerged that further explained work in hotels. These factors were only apparent through the ethnographic work where I was immersed in the OC of hotels and the working lives of associated employees.

Earlier in this chapter, the advantages of triangulation of methods and ethnography were discussed. The reasons why ethnography was chosen as an appropriate method of enquiry for this study were also deliberated. Ethnography was defined as an interpretive and naturalistic inquiry into the shared meanings by which group members make sense of and add importance to everyday behaviour (Dharamsi & Charles, 2011). At an exploratory level of inquiry into the phenomenon of hospitality work, I found that ethnography provides flexibility to explore, describe and analyse the working lives of hospitality workers from within the cultural context of their working environment. This exposition presents rich portrayals of the day-to-day experiences of hospitality workers and the meanings these hold for them.

Fieldwork for this part of the study was based on my time working in the hospitality industry and assuming the role of participant observer. It included semi-structured, face-to-face, in-depth interviews with frontline service personnel, and observation of interactions between both hotel personnel and hotel personnel and guests. Information and vignettes were recorded daily in field notebooks and a personal journal and formed my data. Gathering information about hospitality workers and OC in hotels was definitely facilitated by my working in the industry. This provided the opportunity to be in close and prolonged contact with the staff and to observe first-hand the various elements of their everyday working lives, the OC of hotels and the 'professional culture' of hospitality work. It also allowed me to make use of situations or opportunities that arose to have meaningful conversations and even to conduct interviews. All employees who participated or volunteered information were informed about my study and were happy to participate. During the time of this study I worked in two hotels. Information about the OC of these hotels was largely facilitated by taking advantage of the immediacy of situations. I observed as fully as possible and asked questions as events occurred. Information about other hotels was gathered in a similar fashion from employees who worked in other hotels, from employees the researcher encountered while on training programmes with employees from other hotels (e.g. from the same hotel chain), and from the Technical and Further Education College (TAFE) and from university students. The latter were from a range of disciplines and worked in the hospitality industry to supplement their income while studying. Encounters with people from other hotels was taken advantage of and used as an opportunity to observe, ask questions and/or conduct an interview.

Participant observation was the fundamental and constant method of data gathering. Just as immediate was the shift from participant observer to interviewer. Interviews were important as an opportunity to explore more formal events, activities or issues that the researcher had observed. Interviews were beneficial as they provided time for interviewees to reflect upon and express their views on particular situations. The congenial nature of interviews in this study and the continued presence of the researcher in the field meant that there was opportunity for interviewees to add to their initial responses. This was not unusual and at times a week or more would elapse when the interviewee would raise an issue again with the researcher. It was not uncommon for interviewees at such times to share with the researcher additional information or other ideas in relation to earlier discussions or interviews as they came to mind.

Being there enabled me to listen to workers who volunteered information in a timely and convenient way. Generally, this occurred in informal situations where the workers spoke of their experiences and feelings, as events were happening, and issues surfaced. They would make statements like, 'Here's something for your study' and then talk about an issue or event that was bothering them. This approach afforded easy access for me to participants and to follow-up with in-depth interviews as the opportunity arose. Also, I found that as a consequence of working relationships between me and participants, this approach allowed interviewees to be more relaxed, open and honest with their responses and interviews were largely informal with interviewees speaking freely about work attitudes, OC and the hotel industry. These occasions were a valuable source of rich data that revealed the true nature of hospitality work.

In this ethnographic approach, I was both an insider and an outsider. As an insider, I was part of the situation, feeling and understanding what it was like for hospitality workers. As an outsider and a detached observer, I was separate from the activities, events and issues and viewed the happenings as objects to be studied. This detachment was an easy shift as I was constantly reminded of the objectives of this research and the valuable opportunities afforded by the ethnographic approach that should be used as effectively as possible. In more recent years, especially as I have come to better understand ethnography and related epistemologies, such as autoethnography, I have realised that my PhD could have easily been autoethnographic but at the time, I was new to qualitative ways of thinking and doing and adopting an ethnographic approach was a big step in itself for me. I know that I have come a long way from such a timorous position and am actually now quite adventurous with my epistemological endeavours.

Ethnographic Techniques

While I worked in the hotel industry for just over 10 years, the systematic ethnographic fieldwork relating to my PhD was conducted for two years, from 2002 to 2003, while I was employed in a range of frontline roles (including supervisor) and mainly food and beverage. By conducting an ethnographic study, my aim was to contribute to already existing theories of attitudes in the workplace and OC. Through close interaction with the employees (especially frontline workers), I sought to examine how they perceived the hotel environment, with a focus on their SPs and the hotel's OC. As I was already working in hotels, my formal ethnographic work could not begin until I had ethics approval. Once I received that approval, I commenced the fieldwork. As I learnt to apply ethnographic techniques in the field, I was able to closely observe people in their natural settings, for long periods of time. I was also able to examine events and interactions and access to documents – all things that gave me insights into SPs and OC.

My extended time in hospitality enabled me to work with and observe a lot of different workers and during a range of shifts. Additionally, I was able to focus on and follow selected staff, allowing me to observe developments and changes, especially in thinking and behaviours. This privileged position meant that I was able to stay true to the ethnographic way by adopting an iterative-inductive approach that allowed for consolidation of data from the interviews and conversations into themes.

In 1995, Hammersley and Atkinson talked of ethnographers participating in research through overt or covert means. Earlier, Gold (1958) had identified four categories of participant observer roles: complete participant, participant-as-observer, observer-as-participant and complete observer, with graduating levels of involvement. I was in the first category of 'complete participant', which refers to a high level of involvement; the last role relates to complete detachment. In all cases, the ethnographer is 'collecting whatever data are available' through 'watching what happens, listening to what is said, [and] asking questions' (Hammersley & Atkinson, 1995: 1). O'Reilly (2012: 18) makes reference to Malinowski's emphasis for ethnography being that it allows the researcher 'to record the organisation of the tribe and the anatomy of its culture; to use minute, detailed observations to log the actual details of daily life; to collect ethnographic statements, narratives, utterances as documents of native mentality'. Ethnography is recognised as 'a toolbox of methods, which are integrated into a multifaceted methodological approach' (Reeves *et al.*, 2013: 1367). For me, this product was the PhD, with an aspiration to one day write a book about hospitality work.

Reflecting on My Journey of Learning, Evolution and Metamorphosis

Ethnography offered a solid research framework for exploring and exposing the underlying, entrenched and invisible elements of the working lives of hospitality professionals. Epistemologically, I realised that by working in the industry, and having a lived experience of hospitality work, I was afforded a unique 'insider's view' of the hotel industry and was able to closely observe behaviour and gain opportunity to ask questions relating to my PhD project in ways richer than would be afforded to outsiders. This insider's view allowed me to be a true ethnographer and proved to be a powerful way of obtaining data which served to enrich knowledge about hospitality workers and the nature of hotel work.

Throughout this chapter, I have presented an autoethnographic chronicle of my journey in engaging with the epistemology that is ethnography. Grounded in the paradigm of interpretivism, ethnography presents a way of knowing about people's social worlds. This includes their workplaces and their behaviours and exchanges within such places. For me, the journey of learning about ethnography has shaped my thinking of qualitative research and has enabled me to think seriously about emerging ontologies and epistemologies. Perhaps most telling is the evolution of use of ethnography since I undertook my PhD. In conducting Google Scholar searches, I eventually focused on a search that had 'ethnography' and 'hospitality' in the title. The search produced 16 results, with articles ranging in publication dates from 1997 to 2017. Of interest is the shift in understanding of ethnography. For example, Davids' (2014) article utilises a 'feminist ethnography'. At the time of my PhD, I was unaware of the feminist lens in management and hospitality research and 'ethnography' was contained to its traditional research design. Broader searches reveal such terms as critical ethnography (Thomas, 1993), organisational ethnography (Neyland, 2008), visual ethnography (Pink, 2013), digital ethnography (Pink, 2016) and affective ethnography (Gherardi, 2019). Fetterman (2019: xi) acknowledges that 'in spite of continual changes in the conceptual and methodological frontier, [ethnography] remains a human instrument'.

As Ybema *et al.* (2009) explain, organisational ethnographies have 'a heritage of long standing' with notable works including the writings of Mayo (1933), Whyte (1948), Blau (1955), Dalton (1959) and Goffman (1959). The application of ethnography to organisational studies continues with key scholars reviving interest in such approaches. Among these are John Van Maanen (1979, 1988, 1995) and Julian Orr (1996), whose works I became quite familiar with during my PhD journey. More recently there is Kostera (2007), Neyland (2007) and Silverman (2007). For me, this is invigorating and encouraging as I enjoy many aspects of ethnography: being in the field engaging with organisations in the real world, observing and participating in the authenticity of organisational life, the writing with latitude (e.g. of rich narrative text) to bring the reader into the field, and the opportunity to revitalise and enhance theory.

My growing explorations into the field of ethnography have shown that it is as hard work and of equal merit to the research of positivists. The rigour and discipline are equally part of the ethnographer's responsibilities. For example, the ethnographer needs to be clear and specific of their research process. In addition, there are the ethical responsibilities to ethnographic work that can complicate and add to the responsibilities to the researcher and to the validity and reliability of the work. When I started out on my PhD journey, I had no idea that my immersion in the environment I was working in and that the people who were my colleagues would play such a central role to my research. In hindsight, I am grateful that I engaged a triangulation of methodologies. This approach allowed me to explore both positivists' and interpretivists' worldviews and 'ways of doing'. My PhD journey was richer and more rewarding for having done so. In the end, the ethnography proved to be the most rewarding aspect of the journey. It enabled me to not only observe and converse in the gathering of data, but it also exposed me to a working-world (the world of hospitality) for which I have a lifelong admiration, respect and appreciation. As a bonus, as a long-term participant observer, I had opportunity to build enduring relationships with participants that continue to give me connections with the hospitality world.

In the hotels, I was one of many employees who was still studying at university and so, it was not unusual for people to be curious about what each of us studied. Often conversations would lead to my proffering details of my study. All the while, I participated in normal work activities for my position, blending in with the work environment and never raising questions about my presence from co-workers. In my mind, I was first and foremost a hospitality worker while at the hotel. So, it was not surprising that often most of my colleagues forgot that I was a PhD scholar. The interest in my work was variable with some colleagues finding the study interesting and were eager to contribute more actively. Management was overwhelmingly supportive of my study and afforded me the approval I needed to run the questionnaire and conduct the ethnography.

As I reflect on my own experiences with paradigms, ontologies and epistemologies, a key message from me in this chapter is for prospective and current researchers to not be afraid to embrace other paradigms or alternative epistemologies within their own ontologies. For example, within interpretivism there is opportunity to engage approaches such as ethnography, as I did, but also other methodologies such as autoethnography, case study, grounded theory and action research. Since my PhD, I have experimented with 'photo-elicitation' (Pryce *et al.*, 2014) and collaborative autoethnography (Pryce & Pryce, 2018).

Conclusion

This chapter has shown that there are many different methods available for conducting research into OC. These range from purely quantitative or etic methods, such as use of questionnaires, to purely qualitative or emic methods, such as ethnography and observation techniques. Each of these methods has inherent advantages and disadvantages. Some are perhaps better suited to different aspects of the phenomenon of OC. For example, questionnaires are more practical in situations where a response is required quickly and unobtrusively. They provide breadth of description and allow generalisability. So, they are perhaps more suited at assessing the upper layers of OC, what Schein (1985) would describe as the artifacts and espoused values. Basic assumptions, however, are best researched through qualitative methods. In this chapter we have seen that ethnography is such an approach that provides a deep, rich, 'thick' description and understanding of OC that could not have been gained through quantitative approaches alone.

This multi-method approach can be thought of as adopting a pragmatic choice of methods, especially in terms of the pragmatist's use of methods that best achieve her/his purpose (Morgan, 2014: 7). As such, the pragmatist's focus is more practical and her/his aim is in, 'solving practical problems in the real world' and so focuses on methodology more than ontological and/or epistemological philosophical positions (Kaushik & Walsh, 2019: 259). Pragmatism 'sidesteps the contentious issues of truth and reality' (Feilzer, 2010: 18) and focuses on 'what works' (Tashakkori & Teddlie, 2003). Here, the pragmatist approach is acknowledged, but I remain steadfast in the construction of reality that was afforded through the interpretivist paradigm and the importance of this paradigm to present the phenomena of OC and its influence on the hotel workers. In particular, through the interpretivist lens, I was able to appreciate how the workers attached meanings to their work and workplaces.

Most memorably, I relished engaging with ethnography and found its specific research approach that combines theoretical, analytical, empirical thinking to be of great value for researchers who may be concerned about undertaking research that is rigorous, robust, and valid. I welcomed being a participant observer who explored and sought to understand the organisational world of hospitality workers and who meshed with the daily lives, practices, rituals, interactions, behaviours, values, beliefs and attitudes of those workers. The experience broadened my knowledge of research and expertise as a researcher. While I recommend ethnography, I am also mindful that as more and more scholars engage in discussions of paradigms, ontologies, and epistemologies, PhD researchers will be challenged more extensively to engage with philosophical discussions. The outcome will hopefully lead to richer studies that connect authentically with lived experiences of those we seek to study.

References

Barley, S. (1983) Semiotics and the study of occupational and organisational cultures. *Administrative Sciences Quarterly* 28 (3), 393–413.
Blau, P. (1955) *The Dynamics of Bureaucracy: A Study of Interpersonal Relations in Two Government Agencies*. Chicago: University of Chicago Press.
Bowling, A. (2002) *Research Methods in Health: Investigating Health and Health Services* (2nd edn). Milton Keynes: Open University Press.

Carlson, N.R., Buskist, W. and Martin, G.N. (2000) *Psychology: The Science of Behaviour*. Boston: Allyn & Bacon.

Creswell, J.W. and Creswell, J.D. (2017) *Research Design: Qualitative, Quantitative, and Mixed Methods Approaches*. London: Sage.

Dalton, M. (1959) *Men who Manage*. Chichester: Wiley.

Davids, T. (2014) Trying to be a vulnerable observer: Matters of agency, solidarity and hospitality in feminist ethnography. *Women's Studies International Forum* 43, 50–58.

Davis, T.R. and Luthans, F. (1984) Defining and researching leadership as a behavioral construct: An idiographic approach. *The Journal of Applied Behavioral Science* 20 (3), 237–251.

Dharamsi, S. and Charles, G. (2011) Ethnography: Traditional and criticalist conceptions of a qualitative research method. *Canadian Family Physician* 57(March), 378–379.

Ellis, C., Adams, T. and Bochner, A. (2011) Autoethnography: An overview. *Conventions and Institutions from a Historical Perspective* 36 (4), 273–290.

Feilzer, M.Y. (2010) Doing mixed methods research pragmatically: Implications for the rediscovery of pragmatism as a research paradigm. *Journal of Mixed Methods Research* 4 (1), 6–16.

Fetterman, D. (2019) *Ethnography: Step by Step* (4th edn). London: Sage.

Gherardi, S. (2019) Theorizing affective ethnography for organization studies. *Organization* 26 (6), 741–760.

Goffman, E. (1959) *The Presentation of Self in Everyday Life*. New York: Anchor.

Gold, R. (1958) Roles in sociological fieldwork. *Social Forces* 36 (3), 217–223.

Hammersley, M. and Atkinson, P. (1995) *Ethnography: Principles in Practice* (2nd edn). Abingdon: Routledge.

Hayes, N. (1997) *Doing Qualitative Analysis in Psychology*. Hove: The Psychology Press.

Hofstede, G. (1994) *Uncommon Sense about Organizations: Cases, Studies and Field Observations*. London: Sage.

Kaushik, V. and Walsh, C. (2019) Pragmatism as a research paradigm and its implications for social work research. *Social Sciences* 8, 255–274.

Korulczyk, T., Biela, A. and Blampied, N. (2019) Being more idiographic in the nomothetic world. *Polish Psychological Bulletin* 50 (2), 207–216.

Kostera, M. (2007) *Organisational Ethnography: Methods and Inspirations*. Lund: Studentlitteratur AB.

Lee, A.S. (1991) Integrating positivist and interpretive approaches to organizational research. *Organization Science* 2, 342–365.

Lee-Ross, D. (2000) Development of the service predisposition instrument. *Journal of Managerial Psychology* 15 (2), 148–160.

Mayo, E. (1933) *The Human Problems of Industrial Civilization*. London: Macmillan.

Morgan, D. (2014) Pragmatism as a paradigm for social research. *Qualitative Inquiry* 20 (5), 1–9.

Mullins, L. (1996) *Management and Organizational Behaviour*. London: Pitman.

Neyland, D. (2008) *Organizational Ethnography*. London: Sage.

O'Reilly, K. (2012) *Ethnographic Methods*. Abingdon: Routledge.

Orr, J. (1996) *Talking about Machines: An Ethnography of a Modern Job*. Ithaca: Cornell University Press.

Ott, S.J. (1989) *The Organisational Culture Perspective*. Pacific Grove, CA: Brooks/Cole Publishing.

Patton, M.Q. (1990) *Qualitative Evaluation and Research Methods* (2nd edn). London: Sage.

Pink, S. (2013) *Doing Visual Ethnography*. London: Sage.

Pink, S. (2016) Digital ethnography. In S. Kubitschko and A. Kaun (eds) *Innovative Methods in Media and Communication Research* (pp. 161–165). London: Palgrave Macmillan.

Pryce, J., Chaiechi, T., Ciccotosto, S. and Loban, H. (2014) Use of photo-elicitation to gain insights into the nature of self-managed teams in the academic world. *Proceedings of the ACSPRI Social Science Methodology Conference*, 7–10 December 2014, Sydney, NSW, Australia.

Pryce, J. and Pryce, H. (2018) Utilising collaborative autoethnography in exploring affinity tourism: Insights from experiences in the Field at Gardens by the Bay. In P. Mura and C. Khoo-Lattimore (eds) *Asian Qualitative Research in Tourism: Ontologies, Epistemologies, Methodologies, and Methods* (pp. 205–219). Cham: Springer.

Reeves, S., Peller, J., Goldman, J. and Kitto, S. (2013) Ethnography in qualitative educational research: AMEE Guide No. 80. *Medical Teacher* 35 (8), 1365–1379.

Riley, M. and Dodrill, K. (1992) Hotel workers' orientations to work. *International Journal of Contemporary Hospitality Management* 4 (1), 23–25.

Rousseau, D.M. (1990) Assessing organizational culture: The case for multiple methods. In B. Schneider (ed.) *Organizational Climate and Culture* (pp. 153–192). San Francisco: Jossey Bass.

Schein, E. (1985) *Organizational Culture and Leadership: A Dynamic View*. San Francisco: Jossey-Bass.

Schein, E.H. (1992) *Organizational Culture and Leadership* (2nd edn). San Francisco: Jossey-Bass.

Schultz, M. (1995) *On Studying Organisational Cultures*. Berlin: Walter De Gruyter.

Sechrest, L. and Sidani, S. (1995) Quantitative and qualitative methods: Is there an alternative. *Evaluation and Program Planning* 18 (1), 77–87.

Shamir, B. (1975) A study of working environments and attitudes to work of employees in a number of British hotels. Doctoral thesis, London School of Economics.

Siehl, C. and Martin, J. (1984) The role of symbolic management: How can managers effectively transmit organizational culture? In J.G. Hunt, D.M. Hosking, C.A. Schriesheim and R. Stewart (eds) *Leaders and Managers: International Perspectives on Managerial Behaviour and Leadership* (pp. 227–269). Oxford: Pergamon Press.

Silverman, D. (2007) *A Very Short, Fairly Interesting and Reasonably Cheap Book About Qualitative Research*. London: Sage.

Tashakkori, A. and Teddlie, C. (2003) Issues and dilemmas in teaching research methods courses in social and behavioural sciences: US perspective. *International Journal of Social Research Methodology* 6 (1), 61–77.

Ticehurst, B. and Veal, G.W. (1999) *Business Research Methods: A Managerial Approach*. Boston: Addison Wesley Longman.

Thomas, J. (1993) *Doing Critical Ethnography*. London: Sage.

Tucker, R.W., McCoy, W.J. and Evans, L.C. (1990) Can questionnaires objectively assess organizational culture? *Journal of Managerial Psychology* 5 (4), 4–11.

Van Maanen, J. (1979) The fact of fiction in organizational ethnography. *Administrative Science Quarterly* 24, 539–550.

Van Maanen, J. (1988) *Tales of the Field: On Writing Ethnography*. Chicago: University of Chicago Press.

Van Maanen, J. (1995) *Representation in Ethnography*. London: Sage.

Whyte, W. F. (1948) *Human Relations in the Restaurant Industry*. McGraw-Hill.

Ybema, S., Yanow, D., Wels, H. and Kamsteeg, F.H. (eds) (2009) Studying everyday organizational life. In S. Ybema, D. Yanow, H. Wels and F.H. Kamsteeg (eds) *Organizational Ethnography: Studying the Complexity of Everyday Life* (pp. 1–20). London: Sage.

9 Neo-Tribalism through an Ethnographic Lens: A Critical Theory Approach

Oscar Vorobjovas-Pinta

There were no road signs, nor bright, vivid, pleading resort billboards along the highway. And all-knowing Siri seemed to be lost as well, off the grid, up here in Far North Queensland. As I drove the tiniest rental car up and down Captain Cook Highway, just me and my bright red Hyundai, I had no way to find my destination other than the brief, cryptic text message I received from the resort manager: '3km past Rex Lookout look for a M30 road sign and after 50m turn right'. I felt like I was trying to find Platform 9¾, except I was lost on the beach. Worse, there was barely any phone signal – every traveller's nightmare. As I learned later, my idyllic and remote destination relies on a satellite phone system, and – cut off from the ease of modernity's ubiquitous, unceasing connectivity – I abruptly felt as if I was adrift in time, as well as space. At the end of a journey of almost eight hours from Hobart, Tasmania, I finally reached my destination: the one and only gay and lesbian resort in Australia. As I entered the reception area, I was greeted with a warm salutation: 'Welcome to the family, we've been waiting for you'. Although the tiny building housing the reception desk, and its welcoming occupant, sat right behind the main hotel building towering above it, this space felt distinctly like an inviting gateway to a truly idyllic resort life. As I walked through the reception, and entered the main resort area – the bar, the restaurant, the ocean! – I noticed this space was filled with friendly gay men, all seemingly enjoying perfect relaxation. It felt immediately as though they had known each other for a long time. I soon discovered I was wrong. How do they know each other? I wondered. How did they find this place? Why, of everywhere on Earth, did they travel here?

Extract from my fieldwork journal,
2 September 2014, Far North Queensland, Australia

Introduction

For a long time, tourism has been closely associated with the mass consumption of pre-packaged holidays at sun-and-beach destinations, which have been marketed aggressively by giant tourism operators. Consequently, the positivist approach to tourism market analysis has come to occupy an increasingly dominant position (Vorobjovas-Pinta, 2018a). Nonetheless, positivism has been challenged lately as the predominant basis of constructing tourism knowledge, with critical and reflexive dialogue leading the current debate (Chambers, 2018; Ren *et al.*, 2010; Wilson & Hollinshead, 2015). The demographic shifts driving overall tourism numbers have likewise led to an increased demand for more personalised, bespoke and intimate holidays. To the academic field of tourism studies, consumer preference for this 'niche' form of tourism has marked a departure from the era of modernity, and the embrace of postmodern understandings of society.

Following on from an era of mass tourism, it is easy to imagine the identity and desires of the tourist in a singular and monolithic mode. The tendency of academic tourism studies towards quantitative enquiry has rested comfortably upon many of these assumptions (Andriotis *et al.*, 2008; Prayag *et al.*, 2015). The field has historically made only limited effort to explore tourism as a deeply multifaceted and even conflicted phenomenon, and as a substrate and stage for the many identity categories and performances which characterise us as individuals – be this age, race, gender, class or sexuality (Ateljevic *et al.*, 2007; Wilson *et al.*, 2008; Wilson & Hollinshead, 2015). As the tourism market itself begins to fracture into increasingly niche and targeted products and consumer categorisations (Novelli, 2012), the limitations of the conventional approach to tourism studies have been thrown into ever starker disarray. The movement towards critical tourism studies repudiates the limited perceptivity of traditional approaches, and with it has come a shift in ontological, epistemological, and methodological models, axioms and assumptions (Ateljevic *et al.*, 2007; Ren *et al.*, 2010).

In this chapter, I provide a brief overview of the critical theory paradigm. I then discuss a novel approach for ethnographic enquiry by introducing neo-tribal theory and its application in tourism research. My approach to the theoretical framework (neo-tribalism) and methodology (ethnography) was rather different. This is because neo-tribalism is seen as an aspect of the postmodern paradigm (Fox & Miller, 1996), whereas I applied it within a context of critical ethnography. The core utility of neo-tribalism to tourism studies may be that it centres intangibles such as belonging, social interaction, mobility and membership in the discussion. This chapter derives from my PhD project that examined the emergent and transient culture of a group of gay men congregating at a resort in far north Queensland, Australia. The aim of the research was to enrich the neo-tribal theory by exploring the role space plays in forming temporal communities. I conducted a short-term ethnography at the gay resort, where I lived, volunteered and researched for six weeks, immersing myself into the resort's day-to-day realities. The chapter discusses how I came to recognise that neo-tribalism was relevant to my research. I present a critical and reflective examination of neo-tribalism, highlighting its strengths and noting its limitations. The focal take-away message that I am intending to make through this chapter is that the alignment between paradigms and theories is not always black and white, nor should it be prescriptive. It is the reasoning and iterative dialogue that we engage with that matters.

A Review of the Critical Theory Paradigm

Critical theory is a social philosophy and a research paradigm that emerged from the Institute of Social Research established in 1923 in Frankfurt, Germany (Chambers, 2007). Critical theory (ger. *Kritische Theorie*) was first defined by Max Horkheimer in his 1937 essay 'Traditional and Critical Theory' (ger. *Traditionelle und kritische Theorie*). It emerged as an opposition to positivist social science. It is characterised as 'ideologically oriented inquiry' (Guba, 1990: 23), which arises from a need for human empowerment, personal liberation, and social change (Chambers, 2007, 2018; Creswell & Poth, 2018). As critical theory stands hand-in-hand with the continuous social critique of class-, gender- and race-based constraints, critical theorists determine reality through historically emergent social, political, cultural, economic, ethnic and gendered principles (Chambers, 2007, 2018; Creswell & Poth, 2018; Lincoln & Guba, 2000). Indeed, social justice is the *raison d'être* of the critical theory paradigm, and, affirming that 'the future of humanity depends on the existence today of the critical attitude' (Horkheimer, 1972: 242). Additionally, Horkheimer (1972: 246) contended that a 'critical theory' is one that not only seeks to 'increase knowledge' but it 'must respect' traditional theories 'on which it has for decades exercised a liberating and stimulating influence', as it seeks to realise people's 'emancipation from slavery'. As Horkheimer (1993: 21) explains, critical theory's goal is social transformation and 'has as its object human beings as producers of their own historical form of life'. The interests of critical theorists, hence, extend beyond merely studying and understanding society and related phenomenon; rather, they seek to critique society and to change it (Guba & Lincoln, 1994).

In a fundamental sense, critical theorists utilise an ontological perspective that accepts the existence of an external reality; this reality, however, cannot be comprehended in a complete or unbiased manner (Guba, 1990). Horkheimer advocated for an interdisciplinary approach that built on traditional theory as the objective and explanatory approaches of science to incorporate 'empathetic, subjective, and historical approaches' (Nichols & Allen-Brown, 1996: 1). This stance is evident in the critique by critical theorists towards positivism for the non-interactive and distant relationship between researcher and participant, as the inevitable consequence of its pursuit of knowledge separate from its origins, values and contexts (Guba, 1990). To critical theorists, the context within which the research is conducted is important and so, research associated with critical theory is mindful of the participants, their involvement, and the interactions of participants with researchers (Nichols & Allen-Brown, 1996). Critical theorists propose that the boundaries of possible knowledge are continuously reconstructed through the interactions between a researcher and a participant (Doucet et al., 2010).

Social reality is based upon the understanding that people's perceptions in regard to the same event or circumstance will differ. Hence, by adopting a critical theory paradigm, I was able to interact with the participants and reflect upon how people see the world, and to comprehend the actions, reasons and motivations behind their perceptions and behaviours. The latter implies that in critical theory paradigm we tend to adopt a convenient subjectivist stance, meaning that a researcher and a participant are interactively intertwined, and, as such, it underpins the initial assumption that the outcome of such interaction is a creative product in its most literal sense (Guba, 1990; Lincoln & Guba, 2000). Dialectics, as applied in social constructivism and critical studies, further enables researchers to interpret, compare and contrast multiple such constructions with each other. This might generate one or more constructions in which there is unanimity in reality or representation, and so gesture towards some shared or universal reality that is co-constructed.

A fundamental assumption of critical theory is that the values of the researcher and participants influence the research and through the collaborative nature of critical inquiry are empowered and emancipated. Critical theorists seek to capture the lived experiences of the participants and identify the 'oppressive constraints' and voices of participants to be captured (Doucet et al., 2010). In so doing, critical theory embraces alternative ideological approaches, including participatory inquiry, feminism and critical race theory. This chapter presents another ideological approach, that of neo-tribalism, to be proffered with the paradigm of critical theory.

Neo-tribalism

In my early to mid-twenties, I used to embark on multiple 2 to 20 days-long solo travels around Europe. When I travel, I usually like to keep to myself and I rarely engage in conversations with strangers. However, on several occasions, people I met while on the road became some of the most valued friends I have today. One might think I am very selective or, maybe, there is some theory involved… I was always wondering what the underpinning qualities or, perhaps, peculiarities are that establish this inseparable bond between the two strangers who just met. I first came across the neo-tribal theory when I was writing my research proposal on gay travellers to pursue a PhD degree at the University of Tasmania, Australia. My then supervisor, Associate Professor Anne Hardy, had just published a paper on the bonds that recreational vehicle owners build while on the road (Hardy *et al.*, 2012). Anne and her co-authors used the neo-tribal theory to uncover aspects such as fellowship and facilitation of a certain lifestyle. At that moment I knew I had a theoretical framework for my PhD.

The theoretical framework of neo-tribalism appeared in the academic literature during the transition from modernity to postmodernity, through the work of Michel Maffesoli (Aubert-Gamet & Cova, 1999; Bennett, 1999; Cova & Cova, 2002; e Silva & dos Santos, 2012). Maffesoli, a French professor of Sociology, coined the concept of neo-tribalism in his book *Les Temps des Tribus* in 1988. The English translation *The Time of the Tribes* was published in 1996. To Maffesoli (1996), neo-tribes are heterogeneous fragments, which persist as a hangover from the era of mass consumption. These fragmentary associations of people are conceptualised as fluid 'neo-tribes' rather than strictly bound subcultures, and this distinction forms the core of Maffesoli's theoretical invention. As these postmodern tribal groupings are innately ambiguous, small-scale, and affectual, they do not fit any of the traditional parameters of modern society (Cova & Cova, 2002). I found these considerations of neo-tribalism pertinent to my research because I was aiming to establish how gay travellers who come from disparate walks of life form strong bonds and how certain spaces could facilitate the creation of belonging, connectedness and mutual affinity.

The fragmentation, and hence, individualism of society has been accelerated as a consequence of developments within industry and commerce. It has been argued that the influences of post-industrial work, and the seemingly inexorable progress of deregulation, cultural admixture and increasing trade collectively termed globalisation, has supplanted conventional sociality with 'radical individualism', which then engenders a 'ceaseless quest for personal distinctiveness and autonomy in lifestyle choices' (Arnould & Thompson, 2005: 873). This assertion in turn echoes observations of anomie and moral disorientation during the transition from pre-industrial to industrial society, as most famously contained in Émile Durkheim's *De la Division du Travail Social* (English: The Division of Labour in Society).

Neo-tribes are known to be inherently ephemeral; they exist so long as there is enough drive *(puissance)* for the tribe to exist and perpetuate itself (Maffesoli, 1996). The emergence and disintegration of these neo-tribal groupings depend on the strength of the bonds between the members and their common interests. Even though neo-tribes are heterogeneous in composition and membership (Cova & Cova, 2002), they have this core homogeneous or unitary characteristic stemming from the focus on a shared passion. A neo-tribe must essentially be described as a network of heterogeneous individuals, with regard to age, income, sex, among other examples. As I read these further descriptions of neo-tribalism, I realised the synergies between this paradigm and the research work I was setting out to accomplish.

The tribe is not formed through commonality in any broader identity label; nor could any incidental uniformity in identity among a tribe be excluded either, by definition. This is to say that individuals hailing from different walks of life become linked together through their shared passion and emotion, and this is what distinguishes the neo-tribe. For example, such passion or emotion might be represented through a membership to a certain gym, or support for yet another rising 'X-Factor' star. In the case of my research, individuals were drawn together because they all chose to come to the only LGBTQI+ resort in Australia and they were gay.

Neo-tribal traits are based upon a collective form of identity that stems from common sentiment, rather than any rational process or thought. E Silva and dos Santos (2012) define common sentiment (or shared passion) as the 'linking value' that glues the members of a particular neo-tribe together. The linking value of a commercial or cultural object is, in this sense, proportional in its power to connect the members of a nascent neo-tribe, and to cement their bonds. Members of a particular neo-tribe are themselves the arbiters of this linking value and relationship, and not the providers of products or services; neo-tribes may form around products never intended to promote such social affiliation, or conversely, goods with advertising infused with identity cues and markers may fail to build a tribe-like following (Aubert-Gamet & Cova, 1999). As such, neo-tribal theory has synergies with the narratives explored under the banner of the critical theory paradigm, particularly, as it pertains to the social critique of class-, gender- and race-based constraints. For example, LGBTQI+ neo-tribes, formed by people from disparate walks of life, can be mobilised and collectively express their emotional connectedness by challenging and actively opposing the notions of marginalisation and exclusion in a case of attending LGBTQI+ events (Vorobjovas-Pinta & Hardy, 2021).

Maffesoli describes neo-tribes as less defined by solidity, in the organisational forms which demarcate much of society, than they are a convergent display of lifestyles and associations which contain the tribal ambiance and state of mind (Robards & Bennett, 2011). Consequently, neo-tribes are not limited in the nature of their associations; they may be organised around transient and interim identities, lucrative commodities, labels, brands or locations. The emergence of the – loosely defined – neo-tribal form is an outcome of the collective conquest of a space; this might be, as one example, how a themed cruise ship functions as a temporary substrate for the manifestation of a new and particular neo-tribe (for parallels with cruise ship industry see Weaver, 2011). In the case of my research, the LGBTQI+ resort space was valued as neo-tribal possibility and gay (and male) homosociality made manifest. The resort held abiding value in the memories and imagination of its guests beyond their time at the resort (Vorobjovas-Pinta, 2018a).

Neo-tribalism and tourism

Grounded in sociological inquiry, neo-tribal theory provides rich opportunities for delving into the social aspects of tourism and, hence, exploring the situated qualities. A neo-tribe is defined as a collection of people hailing from different walks of life, united through their shared sentiments, rituals, symbols and kinship (Bennett, 1999; Cova & Cova, 2002; Goulding & Shankar, 2011; Hardy & Robards, 2015). Neo-tribal theory can be used as an essential tool for exploring the sensitivities inherent to a particular group of people, and to their particular destination or place. Hardy and Robards (2015: 444) explain the critical potential neo-tribal theory provides:

> The neo-tribal lens offers a critical departure from traditional segmentation approaches, because rather than focusing on tangible aspects such as their common motivations, demographic characteristics, or travel behaviour once at their destination, it focuses on the intangible aspects, which create a sense of belonging among tourists.

Applications of the neo-tribal concept to tourism studies are rather rare. Explorations of specific consumer categories within tourism have often leaned upon class- or wealth-stratification and have been rigid in their models. This entails the imposition of a rubric where tourist behaviour is atomised into its socio-economic, cultural, and supply-side determinants, but is ultimately regarded as a set of choices made by individual consumers within this larger system (Andriotis et al., 2008; Prayag et al., 2015). The emergent cultures and decision-making of groups is not itself considered. The apparent consensus underpinning this epistemological stance is only now being undermined by the larger shift towards niche tourism, and the necessary reorientation this entails. Tourism research is challenged when the product is no longer a tourist and his or her enjoyment of the destination, in any stable and fixed sense. Under this new paradigm for research, the product becomes the social experience of a new environment as it is shared with like-minded others. It is then impossible to describe, let alone research, tourism behaviour in purely individual terms. To only sort tourism consumers by their demographic categories is to imagine these people as incapable of meaningful association, and of collective behaviour and decision-making (Cova & Cova, 2002; Hardy & Robards, 2015). The core utility of neo-tribalism to tourism studies is that it centres intangibles such as belonging, social interaction and association, mobility and membership in the conversation. To conventional modes of tourism inquiry, these facets are regarded as wholly inaccessible, and are omitted from attempts to sort and categorise tourists. This omission might result in substantial failures in the effort to distinguish one type of consumer from another and creates a set of assumptions – both underlying traditional research methods, and then in turn perpetuated by them – which are insensitive to why some tourists choose to travel, to what end, and with whom.

Despite its rich qualities and clear applicability to tourism as a discipline, at the time of writing the neo-tribal approach had only been

Table 9.1 Applications of neo-tribalism in tourism research

Focus	Authors
Gay tourism	Vorobjovas-Pinta (2017, 2018a, 2018b), Vorobjovas-Pinta and Hardy (2021), Vorobjovas-Pinta and Lewis (2021)
Airbnb	Goodfellow et al. (2018), Hardy et al. (2021)
Clubbing culture	Goulding and Shankar (2011)
Recreational vehicle (RV) users	Hardy et al. (2013), Hardy et al. (2012), Hardy and Robards (2015)
Cruise ships	Weaver (2011), Kriwoken and Hardy (2018)

employed on four occasions in the context of academic tourism and leisure research. Table 9.1 illustrates the five focus areas of these research projects. Hardy et al. (2018) have recently explored further insight into applications of neo-tribal theory in the edited book titled *Neo-Tribes: Consumption, Leisure and Tourism*.

I was encouraged by this literature and became further convinced that neo-tribalism was the theoretical framework that best applied to my research. It did fit well with the critical dialogue I was building.

Understanding the characteristics of neo-tribalism

Neo-tribalism is characterised generally by fluidity, occasional gatherings and dispersal (Bennett, 1999; Goulding & Shankar, 2011; Hardy et al., 2013). I proposed that neo-tribalism could be condensed into four overarching characteristics: shared sentiment, rituals and symbols, fluidity in membership and space (Vorobjovas-Pinta, 2018a). Figure 9.1 depicts these characteristics, and the relationships between them. The space element is highlighted as my PhD research focused on space as a point of coherence around which neo-tribes form.

Shared sentiment within neo-tribal theory represents an impulsive desire to seek out others with shared interests, sensibilities and passions; this then has an impact on patterns of consumption (Cova & Cova, 2002; Hardy & Robards, 2015). Similar patterns of consumption can themselves act as means of connection between people, and the potential for consumer behaviour to express, strengthen and share identity has clear import from a marketing perspective (Aubert-Gamet & Cova, 1999; Cova & Cova, 2002). This aspect has synergies with Turner's (1969) conceptualisation of 'communitas', whereby a sense of fellowship can subsume individual hierarchies of status. The overarching phenomenon demanding consideration is the linkage between shared sentiment and its empowerment of individual members of a neo-tribe to gain strength, and a sense of identity, from their envelopment and connection within that group.

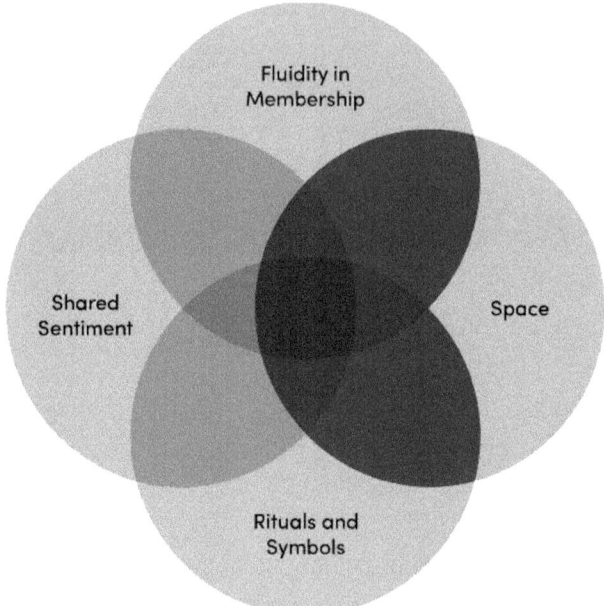

Figure 9.1 Neo-tribal characteristics
Source: Reproduced from Vorobjovas-Pinta, 2017.

Neo-tribal groupings driven by the shared sentiment also possess unique rituals and symbols. Rituals and symbols pertain to the neo-tribe as they help it to shape, sustain and understand the social bonds between the members (Maffesoli, 1996). The need for such rituals and symbols is indeed heightened, and not diminished, by their ephemeral nature; the need to 'consolidate and affirm their union' leaves dynamic neo-tribes in a constant effort to develop appropriate symbols and means of strengthening their unstable bonds (Arnould & Thompson, 2005: 874).

Fluidity in membership is understood as the phenomenon of people's coming together from different walks of life for a shared purpose. The existence of the neo-tribal group is temporary and transitory, in that its removal from day-to-day routine is only enabled by its essential transience. Neo-tribal membership does not propose permanent and structural change. There is no compulsion to even maintain one exclusive tribal affiliation (Goulding & Shankar, 2011). People may not only belong to multiple tribes, but they may also experience paradoxical conflicts within and between them (Cova & Cova, 2002).

Cova and Cova (2002) suggest that the tribe, or at least part of its membership, experiences a need to occupy a defined space for its tribal gatherings. Likewise, Hughson (1999) suggested that neo-tribes gravitate towards a central point of assembly. In this framework, spaces are used as sites of performance for the collective manifestation of identity. Spaces, for this purpose, can be physical (e.g. a football stadium, a concert hall) or virtual (e.g. internet forum, a smartphone app) (Robards & Bennett, 2011).

Neo-tribes will choose their 'points of gathering' where they might 'parade' their 'collective identities' based on their shared tribal traits and collective formations (Hughson, 1999: 14). Such sites of performance are essential for neo-tribes to exist; spaces provide the ground for ideological and tribally aligned cultural manifestation, without which sentiments and rituals cannot be expressed or manifested. In the context of my research, a gay resort for gay visitors is readily understood as a stage for cultural manifestation, and these spaces have been interpreted as the 'anchoring places' that provide a momentary home or temple for a neo-tribe (Aubert-Gamet & Cova, 1999; Vorobjovas-Pinta, 2018b).

While attention to sociocultural interconnectedness and relationality is of course important, these discourses can transmute space into a purely figurative concept; yet it remains productive to consider how spatial delineations and organisations themselves can have social, economic and political consequences, and might productively catalyse and inspire alternative socialities. Spatial division or seclusion is a pre-condition of the conversion of a space into the territory or anchoring place of a neo-tribe, and the ephemeral realisation of the experience of the tribe.

The objective of my PhD thesis was not to delve further into the ontology of space, but to rather extend the applications of neo-tribal theory by developing its spatial characteristic. To this end, it has been noted that the study of the tourism and leisure spaces of LGBT travellers are especially productive, as it permits the researcher to subject 'the "taken-for-granted" to critical interrogation', and so promises wider relevance than LGBT cultures and communities alone (Markwell, 1998: 21–22). To address the paucity of analysis of space as it underpins neo-tribal theory, and by leveraging the possibilities of LGBT space, my thesis interrogated the aforementioned neo-tribal characteristics and their interconnectedness and relations to elucidate the importance of space in the neo-tribal dynamic. My findings suggest that space enables the neo-tribal behaviour through fluidity in membership, shared sentiment, and rituals and symbols (Vorobjovas-Pinta, 2018a, 2018b).

Application of Neo-tribalism and Ethnography

My research aimed to gain a better understanding of the role of space as the point of coherence around which neo-tribes form, and to achieve this through studying the motivations and behaviours of gay travellers. Ethnographic research facilitated the achievement of these aims by promoting a close embeddedness of the researcher in the environs of gay tourism. The assertion has been made that the social world cannot be researched without being part of it, and this stands as a core tenet of this research project (Tedlock, 2003). The results of this work and my reflections of the research process concur with Tedlock's (2003) argument that the application of ethnographic techniques enables a greater understanding of the beliefs, motivations and behaviours of the study participants. I maintain that the awareness of gay traveller's motivations, beliefs and behaviours contributes towards legitimising the spatial characteristic as it exists within neo-tribal theory. From the ethnographic perspective, space provides a fluid, productive, and dynamic basis for engagement (Davies, 2009). These sentiments are elaborated by Vorobjovas-Pinta and Hardy (2014: 641), in that 'neo-tribes may be organised around interim […] places (locations), as the emergence of the neo-tribes is an outcome of the collective conquest of a space'. In the broadest sense, the insights of this project reasserted the potential for ethnographic methods to provide more holistic social accounts, and that these methods are especially useful in establishing linkages between the elements of neo-tribal theory.

Ethnography is both an epistemology and a methodology. As an epistemology, it has been traditionally aligned to the constructivism/interpretivism paradigm (Agar, 2006; Creswell & Poth, 2018). However, the critical approach to ethnography strengthens it as a reflexive epistemology, whereby it becomes more engaging and useful (Foley, 2002). The ways in which we, scholars, engage with individual ethnographic research accounts rests on various theoretical and disciplinary perspectives navigating our logic-in-use (Green *et al.*, 2012). On an epistemological level, I took a subjectivist stance as this study required close communication with the participants. Such communication allowed me to maintain a close proximity to the participants' perceptions of their own set of circumstances (Creswell & Poth, 2018). As the gay resort visitors related their stories and their understandings of the world, my task was to interpret the acquired information and to assess their broader conceptual insights and linkages. Epistemologically, the aim of my research was to gather rich and meaningful insights into the everyday lives of gay resort visitors, as well as to empower them to critically reflect upon their lived experiences of the resort. As such, the information obtained from the study participants can be understood as an outcome of my interactions with them. The subjectivity inherent to this process assisted in inductively building upon existing knowledge of neo-tribal theory.

Ethnography as a methodology is a research approach and process, where a researcher interacts and/or observes a study's participants in a real-life setting and produces a detail and accurate descriptions of such accounts (Creswell & Poth; 2018; Lune & Berg, 2017). A critical ethnographic approach provides research methods and techniques for the analysis of cultural practises, and for the depiction of social change and of the processes by which particular social phenomena are constructed (Creswell & Poth, 2018; Lune & Berg, 2017). LeCompte and Schensul (2010) outline the following ethnographic data collection methods: observations, tests and measures, surveys, interviews, content analysis, elicitation methods, audio-visual methods, spatial mapping and network research. I have used observations and interviews in my research. I have aligned the hermeneutic approach with the research questions, and thus this has allowed me to depict the views of individual gay travellers accurately. Conversely, the dialectic extension of this, facilitated comparison and contrast of these various constructions with one another. The latter facet was significant to this study, as it functioned as a tool for the assessment of possible differences and similarities between gay travellers' behaviours, motivations, and decision-making processes, as applied to their shared space.

Ethnographic research differs from other types of qualitative inquiry in the degree to which the researcher becomes immersed in the fieldwork; this fieldwork being the very reality of the group researched. Despite this positionality, several authors argue that the researcher exists both within and around a particular field (Davies, 2009; Tedlock, 2003). It has been suggested that the social world cannot be researched without being part of it (Tedlock, 2003), as such I, as a researcher, could not separate myself from knowledge that I accumulated prior to arriving to the resort – my fieldwork place. My arrival and presence in the field meant that I have stepped out of my day-to-day space. Tedlock (2003) suggests that a researcher, who is an insider, possesses an advantage in terms of gathering unequivocal facts and data, which then contribute towards gaining in-depth insights and access into daily routines. As such, as a gay man myself I was able to act as an insider researcher, and I believe that this has ensured a more authentic and productive form of research. I was occupying the dual role of researcher and employee, and, as such, I was straddling a spatial dichotomy. First, I had to maintain deep levels of immersion while not allowing myself to be drawn too fully into the 'resort lifestyle', and second, I had to remain peripheral to the centre of attention while still ensuring the 'thickness' of data collected (Vorobjovas-Pinta & Robards, 2017).

Both the non-intrusive nature of the research and its limited time frame led me to employ the following methods: participant observation – including non-formal 'chat-type' conversations – and in-depth interviews. I adopted an ontological stance, which holds that 'people's knowledge, views, understanding, interpretations, experiences, and interactions are meaningful properties of the social reality' (Mason, 2010: 63). The epistemological stance is built subsequently upon these ontological properties, which are themselves constructed through the direct and active interaction between the inquirer and the participant (Mason, 2010). A more detailed discussion on the research methodology, and the value of insider ethnography, can be found in Vorobjovas-Pinta and Robards (2017).

There arguably exists a reciprocal relationship between the neo-tribal theoretical perspective and qualitative inquiry. It has been suggested that neo-tribal theory necessitates a qualitative approach, as the theory explicitly concerns sensitive phenomena, and a quantitative exploration of these phenomena might lead only to an inadequacy of credible evidence of what is intended to be explored (Greenacre *et al.*, 2013). As Greenacre *et al.* (2013: 954) note:

> It is questionable whether a quantitative approach to the application of [...] [neo-tribal theory] is able to detect sensitive phenomenon, such as temporal order in community formation. This need for sensitivity presents an interesting challenge for methodological selection. The choice of a quantitative method can result in great difficulty in assessing the absence of sensitive phenomena, as they simply won't appear in the results.

My reasoning behind selecting qualitative research methods for this study was to gain a 'more authentic' and 'less exploitative' picture (Greenacre *et al.*, 2013), which is backed by deep insights into people's sensitive behaviours and attitudes, as opposed to broader generalisations (Hughes *et al.*, 2010). Moreover, the impetus for adopting a qualitative and not quantitative approach emerges from the evident truth that '[...] if there is one thing, which distinguishes humans from the natural world, it is their ability to talk. It is only by talking to people, [...] that we can find out what they are thinking, and understanding their thoughts goes a long way towards explaining their actions' (Myers, 2013: 6).

By adhering to the 'embeddedness' principle of ethnographic research, the methods chosen enabled me to gather empirical insights into social routines potentially inaccessible to the general public. Ethnographic methods not only then yield substantial and insightful knowledge of gay travellers' behaviours, but the ethnographic approach further provides a particular thickness, and richness of detail of the processes in their contexts. This represents a chance to understand how particular spaces affect the motivations, behaviours and beliefs of gay travellers, and the inverse, in how these behaviours, motivations and beliefs shape and delineate spaces themselves. The argument for ethnographic methods was sustained for my research project, as they match closely its aims to better comprehend whichever aspects legitimise the spatial characteristic of neo-tribal theory.

Reflections on the Application of Neo-tribal Theory

I built my analysis upon the argument by Hughson (1999: 14), who affirmed that 'an awareness of the social and cultural geography of relevant spaces is [...] crucial to the study of neo-tribes'. I posed the argument that there is a lack of ethnographic studies pertaining to gay tourism, and a specific absence of understanding of exclusively gay tourism spaces, such as gay resorts. The critical aspect of my ethnography was critical in the sense that I did not only attempt to criticise society in their treatment of LGBTQI+ communities, but, more importantly, I challenged the existing status quo in order to help society to transform for the better. My methodological approach was shaped by my own position as a gay man, alongside my cultural and social upbringing. I shared a similar worldview with the participants of my study, including social norms and expectations, and this similarity ensured my native-like immersion into the set environment. This resulted in my PhD research being more reflexive, perceptive and authentic, where participants were permitted to express their morality without implied societal judgement or prejudice.

By focusing upon everyday holiday experiences of gay resort visitors, I analysed the intimate engagement and interplay between the resort space and the resort's patrons. The main argument and contribution of my thesis was that spaces are the sites of performance where collective identities and cultures are manifested. Simultaneously, they are themselves formed by and formative of these identities, unstable spatial delineations re-align with unstable identity boundaries, and vice versa. I asserted further that the fleeting nature of neo-tribal membership is itself enabling. It depicts a transient and transitory escape from day-to-day routines. The space not only mediates, but produces and governs the rituals and symbols, fluid membership and shared sentiments which underlie neo-tribal experience. Ultimately, space is represented as the linking value, which makes these other characteristics legible and real.

Furthermore, I sought to record the contours of the neo-tribe found at the resort, to relate this to my broader theoretical formulation, and to determine the mechanism and process of neo-tribal dissipation, in its distinctive quality, membership and constructed reality. I examined the collective behaviour of gay resort visitors and capitalised upon the enclosed resort space as an accessible microcosm of dynamic neo-tribal affiliations and gatherings. The gay resort contained meaning to its clientele as an imagined place of return, and an idyll always accessible to gay men from their 'real' lives. Consequently, resort visitors could entrench the divide between their resort lives and their 'real' selves, and then draw on this division as a source of strength and solidarity outside the resort space. The resort becomes an imaginary once departed but gives its patrons a continuing sense of belonging and continuity through its reification of a gay social ideal. The neo-tribe of the resort granted its members ownership of territory, and with it, permission to imagine alternative and transcendent socialities. I established that space is elevated in its relationship to neo-tribes, and that feelings of collective ownership by the tribe of their space are constitutive of tribal identity itself; entry into and shared ownership of tribal space is in some sense the currency of tribal membership and belonging. The overarching argument was that as space becomes the linking value of a neo-tribe, the tribe dissipates and endures no longer once this shared environment is taken away. This finding makes a major contribution to the understanding of neo-tribal theory in the academic literature.

Conclusion

In this chapter, I looked at the neo-tribal theory through a lens of the critical theory paradigm, adopting an ethnographic approach, and highlighted the argument that the alignment between theories and paradigms is not always black and white. It is much more nuanced, and indeed it is the iterative dialogue that matters. I presented the evolution of neo-tribal theory, and its arrival in the academic tourism and leisure literature. The discussion presented provides the breadth of conceptualisations of the theory, with particular concern for the area of tourism. My reinterpretation of the theory was detailed through the introduction of four overarching characteristics defining neo-tribalism: (a) fluidity in membership; (b) rituals and symbols; (c) shared sentiment; and (d) space. The analysis of the theoretical underpinnings of neo-tribalism reveals the lack of inquiry into space as the characteristic enabling neo-tribal coalescence.

Fundamentally, my research not only addressed the gap left by the dearth of ethnographic inquiry into gay tourism, but also contributed to neo-tribal theory by showcasing that the lived experience of ephemeral groupings – neo-tribes – can be captured through the use of short-term ethnography. The practical implications of my research suggested that there is an opportunity for tourism operators to encourage customer loyalty and repeat business through the deliberate cultivation of neo-tribal characteristics, as enacted through the deliberate manipulation of space.

I truly hope that you, as a reader, feel that you have gained a genuine and deeper understanding of the neo-tribal theory and its intersection with the critical theory paradigm and ethnographic research methods. I hope too that you take neo-tribalism with you, as a powerful tool for the observation and understanding of how people build collective meaning in a complex and dynamic world. Neo-tribalism is a rejection of social atomisation in societies where – on first glance – it seems we have no time to find meaning in each other, and where the inner experiences of those hailing from different walks of life are inaccessible to us. Neo-tribes use space as a means of re-establishing ownership, solidity, and locality, and to share these blessings and dividends with people different from ourselves; here, space becomes the very fulcrum of neo-tribal coalescence.

Acknowledgement

Parts of this chapter appeared in 'Gay neo-tribes: an exploration of space and travel behaviour' (Doctoral dissertation) by Oskaras Vorobjovas-Pinta, University of Tasmania, Sandy Bay, Australia, 2017.

References

Agar, M. (2006) An ethnography by any other name... *Forum Qualitative Sozialforschung/ Forum: Qualitative* 7 (4). http://www.qualitative-research.net/index.php/fqs/article/view/177/396

Andriotis, K., Agiomirgianakis, G. and Mihiotis, A. (2008) Measuring tourist satisfaction: A factor-cluster segmentation approach. *Journal of Vacation Marketing* 14 (3), 221–235.

Ateljevic, I., Morgan, N. and Pritchard, A. (2007) *The Critical Turn in Tourism Studies: Innovative Research Methodologies.* Harlow: Elsevier.

Arnould, E.J. and Thompson, C.J. (2005) Consumer culture theory (CCT): Twenty years of research. *Journal of Consumer Research* 31 (4), 868–882.

Aubert-Gamet, V. and Cova, B. (1999) Servicescapes: From modern non-places to postmodern common places. *Journal of Business Research* 44 (1), 37–45.

Bennett, A. (1999) Subcultures or neo-tribes? Rethinking the relationship between youth, style and musical taste. *Sociology* 33 (3), 599–617.

Chambers, D. (2007) Interrogating the 'critical' in critical approaches to tourism research. In I. Ateljevic, N. Morgan and A. Pritchard (eds) *The Critical Turn in Tourism Studies: Innovative Research Methodologies* (pp. 105–120). Oxford: Elsevier.

Chambers, D. (2018) Tourism research: Beyond the imitation game. *Tourism Management Perspectives* 25, 193–195.

Cova, B. and Cova, V. (2002) Tribal marketing: The tribalisation of society and its impact on the conduct of marketing. *European Journal of Marketing* 36 (5-6), 595–620.

Creswell, J.W. and Poth, C.N. (2018) *Qualitative Inquiry and Research Design: Choosing among Five Approaches*. Thousand Oaks: Sage Publications.

Davies, A.D. (2009) Ethnography, space and politics: Interrogating the process of protest in the Tibetan freedom movement. *Area* 41 (1), 19–25.

Doucet, S.A., Letourneau, N.L. and Stoppard, J.M. (2010) Contemporary paradigms for research related to women's mental health. *Health Care for Women International* 31 (4), 296–312.

e Silva, S.C. and dos Santos, M.C. (2012) How to capitalise on a tribe. *The Marketing Review* 12 (4), 417–434.

Foley, D. (2002) Critical ethnography: The reflexive turn. *International Journal of Qualitative Studies in Education* 15 (4), 469–490.

Fox, C.J. and Miller, H.T. (1996) Round #1: What do we mean when we say "real" in Public Affairs: The modern/postmodern distinction. *Administrative Theory and Praxis* 18 (1), 101–108.

Goodfellow, D.L., Hardy, A. and Dolnicar, S. (2018) Communication-regulated social systems. In S. Dolnicar (ed.) *Peer-to-Peer Accommodation Networks: Pushing the Boundaries* (pp. 225–234). Oxford: Goodfellow Publishers.

Goulding, C. and Shankar, A. (2011) Club culture, neotribalism and ritualised behaviour. *Annals of Tourism Research* 38 (4), 1435–1453.

Green, J., Skukauskaite, A. and Baker, B. (2012) Ethnography as epistemology: An introduction to educational ethnography. In J. Arthur, M.I. Waring, R. Coe and L.V. Hedges (eds) *Research Methodologies and Methods in Education*. London: Sage Publications.

Greenacre, L., Freeman, L. and Donald, M. (2013) Contrasting social network and tribal theories: An applied perspective. *Journal of Business Research* 66 (7), 948–954.

Guba, E.G. (1990) The alternative paradigm dialog. In E.G. Guba (ed) *The Paradigm Dialog* (pp. 17–30). Newbury Park: Sage Publications.

Guba, E. and Lincoln, Y. (1994) Competing paradigms in qualitative research. In N. Denzin and Y. Lincoln (eds) *Handbook of Qualitative Research* (pp. 105–117). Thousand Oaks, CA: Sage.

Hardy, A., Bennett, A. and Robards, B. (eds) (2018) *Neo-Tribes: Consumption, Leisure and Tourism*. Cham: Palgrave Macmillan.

Hardy, A., Dolnicar, S. and Vorobjovas-Pinta, O. (2021) The formation and functioning of the Airbnb neo-tribe. Exploring peer-to-peer accommodation host groups, *Tourism Management Perspectives* 37 (January 2021), 100760.

Hardy, A., Gretzel, U. and Hanson, D. (2013) Travelling neo-tribes: Conceptualising recreational vehicle users. *Journal of Tourism and Cultural Change* 11 (1-2), 48–60.

Hardy, A., Hanson, D. and Gretzel, U. (2012) Online representations of RVing neo-tribes in the USA and Australia. *Journal of Tourism and Cultural Change* 10 (3), 219–232.

Hardy, A. and Robards, B. (2015) The ties that bind: Exploring the relevance of neotribal theory to tourism. *Tourism Analysis* 20 (4), 443–454.

Horkheimer, M. (1972) *Traditional and Critical Theory*. New York: Herder and Herder.
Horkheimer, M. (1993) *Between Philosophy and Social Science*. Cambridge, MA: MIT Press.
Hughes, H.L., Monterrubio, J.C. and Miller, A. (2010) 'Gay' tourists and host community attitudes. *International Journal of Tourism Research* 12 (6), 774–786.
Hughson, J. (1999) A tale of two tribes: Expressive fandom in Australian soccer's A-league. *Culture, Sport, Society* 2 (3), 10–30.
Kriwoken, L. and Hardy, A. (2018) Neo-tribes and Antarctic expedition cruise ship tourists. *Annals of Leisure Research* 21 (2), 161–177.
LeCompte, M.D. and Schensul, J.J. (2010) *Designing and Conducting Ethnographic Research: An Introduction*. Plymouth: AltaMira Press.
Lincoln, Y.S. and Guba, E.G. (2000) Paradigmatic controversies, contradictions, and emerging confluences. In N.K. Denzin and Y.S. Lincoln (eds) *Handbook of Qualitative Research* (2nd edn, pp. 163–188). Thousand Oaks: Sage Publications.
Lune, H. and Berg, B.L. (2017) *Qualitative Research Methods for the Social Sciences*. Harlow: Pearson.
Maffesoli, M. (1996) *The Time of the Tribes*. London: Sage Publications.
Markwell, K. (1998) Space and place in gay men's leisure. *Annals of Leisure Research* 1 (1), 19–36.
Mason, J. (2010) *Qualitative Researching*. London: Sage Publications.
Myers, M.D. (2013) *Qualitative Research in Business and Management*. London: Sage Publications.
Novelli, M. (ed.) (2012) *Niche Tourism*. Oxford: Routledge.
Nichols, R.G. and Allen-Brown, V. (1996) Critical theory and educational technology. In D. Jonassen (ed.) *Handbook of Research for Educational Communications and Technology* (pp. 1–29). New York: Simon & Shuster Macmillan.
Prayag, G., Disegna, M., Cohen, S.A. and Yan, H. (2015) Segmenting markets by bagged clustering: Young Chinese travelers to Western Europe. *Journal of Travel Research* 54 (2), 234–250.
Ren, C., Pritchard, A. and Morgan, N. (2010) Constructing tourism research: A critical inquiry. *Annals of Tourism Research* 37 (4), 885–904.
Robards, B. and Bennett, A. (2011) MyTribe: Post-subcultural manifestations of belonging on social network sites. *Sociology* 45 (2), 303–317.
Tedlock, B. (2003) Ethnography and ethnographic representation. In N.K. Denzin and Y.S. Lincoln (eds) *Handbook of Qualitative Research* (pp. 165–213). Thousand Oaks: Sage Publications.
Turner, V.W. (1969) *The Ritual Process: Structure and Anti-structure*. Chicago: Aldine Publishing.
Vorobjovas-Pinta, O. (2017) Gay neo-tribes: An exploration of space and travel behaviour. Doctoral dissertation, University of Tasmania, Sandy Bay, Australia.
Vorobjovas-Pinta, O. (2018a) Gay neo-tribes: Exploration of travel behaviour and space. *Annals of Tourism Research* (72), 1–10.
Vorobjovas-Pinta, O. (2018b) 'It's been nice, but we're going back to our lives': Neo-tribalism and the role of space in a gay resort. In A. Hardy, A. Bennett and B. Robards (eds) *Neo-Tribes: Consumption, Leisure and Tourism* (pp. 71–87). Cham: Palgrave Macmillan.
Vorobjovas-Pinta, O. and Hardy, A. (2014) Rethinking gay tourism: A review of literature. In P.M Chien (ed.) *Proceedings of CAUTHE 2014: Tourism and Hospitality in the Contemporary World: Trends, Changes and Complexity* (pp. 635–644). Brisbane: The University of Queensland.
Vorobjovas-Pinta, O. and Hardy, A. (2016) The evolution of gay travel research. *International Journal of Tourism Research* 18 (4), 409–416.
Vorobjovas-Pinta, O. and Hardy, A. (2021) Resisting marginalisation and reconstituting space through LGBTQI+ events. *Journal of Sustainable Tourism* 29 (2–3), 447–465.

Vorobjovas-Pinta, O. and Robards, B. (2017) The shared oasis: An insider ethnographic account of a gay resort. *Tourist Studies* 17 (4), 369–387.

Vorobjovas-Pinta, O. and Lewis, C. (2021) The coalescence of the LGBTQI+ neo-tribes during the pride events. In C. Pforr, R. Dowling and M. Volgger (eds) *Consumer Tribes in Tourism: Contemporary Perspectives on Special-interest Tourism* (pp. 69–81). Singapore: Springer.

Weaver, A. (2011) The fragmentation of markets, neo-tribes, nostalgia, and the culture of celebrity: The rise of themed cruises. *Journal of Hospitality and Tourism Management* 18 (1), 54–60.

Wilson, E., Harris, C. and Small, J. (2008) Furthering critical approaches in tourism and hospitality studies: Perspectives from Australia and New Zealand. *Journal of Hospitality and Tourism Management* 15 (1), 15–18.

Wilson, E. and Hollinshead, K. (2015) Qualitative tourism research: Opportunities in the emergent soft sciences. *Annals of Tourism Research* 54, 30–47.

10 Navigating the Complex Variety of Feminisms

Linda Colley and Sue Williamson

Introduction

Feminism can be a fraught term and studying feminist theories requires consideration of the waxing and waning of support for feminism and feminist activism. Strong feminist gains were made in the 1970s and 1980s, followed by some backlash against those gains in the 1990s – including from some feminists (see, for example, Roiphe, 1993). Some even claimed we had entered a post-feminist era, although as with much feminist theory, this term has multiple meanings (Gill, 2017). More recently, scholars have discussed how:

> Feminism is a word to shun, heavy, soiled and laden with baggage, a word difficult to claim. … [with a] metaphorical weight that is the anathema of the shining, white, upwardly mobile neoliberal girl subject. (McRobbie, 2009, cited in McKnight, 2016: 6)

While there has been some resurgence in discussion of feminism in recent years, and indeed a focus on women's safety and rights epitomised by the #metoo movement, there has also been a backlash or counter-movement in the form of #notallmen, where individual men reject responsibility for the violence perpetrated by some men (Ford, 18 June 2018).

To understand feminist methodology, it is essential to start with understanding these developments in thinking and theorising around feminism. We begin by tracing the evolution of feminism from the first-wave focus on the right to vote and own property, through the second-wave focus on equality at home and at work and its many subsets, to the third wave of feminism which focused on individual women and included neoliberal feminism, to the emerging fourth wave of feminism, sparked by #metoo. We then outline how we applied the paradigm to both our method and data analysis, and how it is essential for in-depth analysis of the barriers to gender equality. We conclude with some of our personal hints on how to enter into the literature and method.

Review of the Paradigm

The authors have different backgrounds and paths to feminist studies. While they both studied industrial relations and human resource management, Sue had specialised in Women's Studies since her undergraduate degree and was more familiar with the trajectory and how organisations are gendered. Linda entered the literature in the late 1990s, but it was not until her PhD studies that she discovered the idea of waves of feminism, which helped to put it into historical perspective.

Waves of feminism

While there are hints of feminist argument in early Greek and Chinese civilisations, it is more usual to trace feminist debates and theory to the late 18th century when feminist theorists such as Mary Wollstonecraft began to demand equal rights as a theoretical argument, and similar debates arose in China and India (Hawkesworth & Disch, 2016: 1). Hawkesworth and Disch (2016: 1) note that:

> Feminism would not exist as a theoretical endeavor without the political struggles for women's empowerment that have emerged in all regions of the world. Grounded in the investigation of women's and men's lives and convinced of the arbitrariness of exclusion based on sexual difference, feminist theory has flourished as a mode of critical theory that illuminates the limitations of popular assumptions about sex, race, sexuality and gender and offers insights into the social production of complex hierarchies of difference.

Feminist theory expanded from the 1970s, in response to a surge in feminism from the 1960s, and with complex inter-disciplinarity across social and natural sciences. The terms *feminism* and *feminist* have been historically and culturally bounded depending on the time and place and speaker (Calvini-Lefebvre *et al.*, 2010; Moses, 2012). The evolution of feminism is often described in waves. While Schein (2014) and others critique this static chronology, it remains a useful starting point for an early career researcher to think about the overall journey. The *first wave* of feminism in the 19th and early 20th century focused on women's rights, including property rights and suffrage and political candidacy.

Most people are more familiar with the *second wave* of feminism, which is often seen as the collective movement from the 1960s to the 1980s, focused on inequalities, including workplace rights, sexual liberation, rights concerning the family and marriage/defacto relationships, and other legal inequalities. Second-wave feminism was underpinned by different ideologies emanating from the women's movements, and consisted of radical feminism, which holds that women are oppressed due to sexual subjugation (Curthoys, 1988: 82). In this paradigm, the only way to achieve women's equality is by overthrowing patriarchy and establishing new systems that are not dependent on women's sexual and reproductive labour. Completely new gender relations would be established (Alvesson & Skóldberg, 2018). The other main stream of the second-wave women's movement was liberal feminism, which adopted a moderate approach. Adherents believed that women's – or gender – equality is to be achieved by working within the institutional structures. In Australia, this gave rise to a particularly Australian variant of feminism, that of the 'femocrat', where feminists worked in the public sector to progress equality for women (Sawer, 2014). Over time, these movements fractured, developed and morphed into a variety of feminisms.

The *third wave* of feminism is more commonly seen as the 1990s to early 2000s, which embraced individualism and diversity rather than the previous collective approach to address systemic issues. Third-wave feminism has also been seen to be 'post-feminist', symbolised by 'girl power', where young women are independent, confident and not bound by the strictures set by previous generations of feminists (Church Gibson, 2004: 139). Some now argue that a *fourth wave* of feminism has emerged since 2008, focusing on sexual harassment, assault and safety, as well as misogyny and equality of outcomes. This is evident in the #metoo movement. It is also evident in expanding international commitment to gender equality, including in the United Nations Sustainable Development Goals (UN, 2018). Linda only discovered the nuances of these third and fourth waves of feminism in the last few years, and this has helped her to make sense of events such as leading political figures such as former Foreign Minister Julie Bishop refusing to be labelled as a feminist (Colley & White, 2019).

These waves of feminism are chronologically cast and supersede the norms, values and goals of the preceding wave (Colley & White, 2019). By the early 1980s there were critiques of the *waves* approach as being too simplistic and linear, rather than a series of contests and relationships (Hemmings, 2005) that overlapped. Further, there were critiques that feminism was not reflecting the experience of all women but rather privileged Caucasian able-bodied women, making it similar to the homogenising tendencies of earlier sociological theories in which all women were invisible (Alvesson & Skóldberg, 2018; Denis, 2008; Hawkesworth & Disch, 2016). This saw the emergence of identity politics (such as black feminism and lesbian feminism). There was also a linking of feminist theory to other theoretical frameworks, such as Marxist feminism or liberal feminism, critiquing the assumptions in other sociological theories that men's experience was the norm and making women's experiences invisible (Denis, 2008). Hawkesworth and Disch (2016: 3) argue that this hyphenation model 'was useful in showing the continuities and shared assumptions' but at the same time 'tended to make feminist theory seem derivative of mainstream schools of thought, and its emphasis on divisions obscured what feminist theories had in common'.

Many scholars believe that the foray into identity politics or hyphenated theories has been a distraction from the real mission, which should have been redistributive feminism. Hemmings (2005: 116) notes that these shifts might be seen as progress beyond categories, or might be seen as a move away from 'the politicised, unified early second wave, through an entry into the academy in the 1980s, and thence a fragmentation into multiple feminisms and individual careers' and an accompanying 'loss of commitment to social and political change'. Denis (2008: 677–678) argues that despite this diverse array of feminisms, there are common assumptions, including that women are legitimate objects of study that are largely socially constructed and often subordinated to men – these common assumptions mean feminism is most often activist rather than neutral. Alvesson and Skóldberg (2018) agree that gender is socially constructed and therefore can change, also giving credence to the underpinning of an activist basis.

Notwithstanding the criticism, we both found the waves of feminism to be a useful heuristic to think about the field. We now provide a summary of recent developments and ways that we found useful to look at feminisms.

Recent developments and ways of looking at feminisms
Neoliberal feminism

From the 1990s, feminism developed an image problem, and was used as a pejorative term. Some scholars suggest that this was a deliberate conservative push to undermine early feminist gains (Faludi, 1991; McRobbie, 2009). Fraser (2009: 4) has argued that second-wave feminism laid the groundwork for a hybrid of contemporary feminism and neoliberalism, being 'drawn into the orbit of identity politics' and being distracted from redistributive justice and a materialist critique to focus on recognition and difference, thereby letting in neoliberalism. Rottenberg (2014: 421) does not agree with all aspects of Fraser's argument, but does agree that neoliberalism has hijacked the feminist agenda. This colonisation of feminism sought to extend neoliberal and individualistic domains, and transfer feminism from a group struggle or social justice movement to an individual responsibility, reducing or removing the role of state feminism. She appreciates Fraser's identification of a contemporary mode of feminism profoundly informed by a market rationality. Indeed, Rottenberg questions whether neoliberal feminism can be understood as just one more domain that neoliberalism has colonised, hollowing out liberal feminism and entrenching neoliberal rationality.

We see this type of argument put forward by high-powered women, such as Facebook executive Sheryl Sandberg, who recommended women lean in to their careers (Sandberg & Scovell, 2015), and by leading Australian conservative politicians such as Julie Bishop. As Rottenberg

(2014) notes, these reflections from high-powered women move away from feminism being about social justice and equity, to being about individual drive and different behaviours from women. We have found this stream of feminist theory particularly useful, as it also informs the different approaches organisations might take when addressing gender inequality – whether they will look for systemic problems and collective solutions, or whether they seek to address gender inequality one person at a time.

Gender mainstreaming and diversity management

The strong gains in collective women's rights, particularly in their entry to and progress within the labour market, were perhaps too threatening to society and led to a recasting of feminism. Gender mainstreaming policies arose as a complement to these strong central policies, intended to make visible the gender assumptions in policies, processes and outcomes, but in practice can remove the focus on gender and require it to 'fight' for attention (Walby, 2005). Walby highlights the inherent tensions within gender mainstreaming: pursuing the goals of gender equality in its own right, and to make mainstream policies more effective by the inclusion of gender analysis can be at odds with mainstream policies, such as those to foster economic competitiveness, and which may rely on women in low-paid work.

Diversity management policies arose as an alternative to stronger and more contentious affirmative action policies. While affirmative action focused on systemic issues and a general pursuit of gender equality as a moral imperative, diversity management policies are more likely based on a business case rationale and focused on individual rather than systemic issues (Strachan *et al.*, 2007). Managing diversity focuses on celebrating the differences between individuals and aims for cultural transformation through 'mainstreaming' equality, so that equality is considered in all policies and strategies, not just in equality policies. Unlike anti-discrimination legislation, it is positive in nature rather than punitive (Kirton & Greene, 2015: 127). A diversity management approach has, however, been criticised for being ahistorical, ignoring power relations and obscuring systemic discrimination (Noon, 2007). We both consider it a form of tacit resistance to acknowledging gender inequality.

Intersectionality

Intersectional analysis is an important recent theoretical contribution to feminist theory. This approach acknowledges that some people face multiple sources of disadvantage or discrimination – for example, a combination of sex discrimination and age discrimination. It addresses some of the fragmentation issues raised by identity politics, allowing them to come back together and focus on the goal of redistributive feminism. Denis (2008) argued that intersectionality arose from the critique within feminist theory. As with 'feminism', 'intersectionality' is a contested term, with theorists debating the boundaries of the concept, the subjectivity of applying an intersectional approach and how it translates into research methodologies (Cho *et al.*, 2013). Regardless of these conceptual difficulties, intersectionality is an important development within social science and feminist research. Linda has found it particularly useful in her research on gender and age in public services, where women accumulate disadvantage across the life course and older women are particularly disadvantaged by the combination of ageism and sexism.

Inequality regimes

Like most theoretical frameworks, organisational theories were also written as if organisational structures and culture were gender neutral. In her ground-breaking work, Acker develops the 'glass ceiling' metaphor (about barriers to women reaching the higher levels in organisations) into a more accurate concept of inequality regimes that obstruct women's opportunities at all levels (Acker, 2006, 2009). Inequality regimes include the:

> systematic disparities between participants in power and control over goals, resources, and outcomes; in work place decision-making such as how to organize work; in opportunities for promotion and interesting work; in security in employment and benefits; in pay and other monetary rewards; and in respect and pleasures in work and work relations. Organizations vary in the degree to which these disparities are present and in how severe they are. (Acker, 2009: 202)

By going beyond gender regimes, Acker's concept incorporates race and class as well as gender into the analysis. This concept provides both a theoretical framing and an analytic approach to gender in organisations. We are big fans of Acker's work, and each had 'aha' moments when we discovered her concept of inequality regimes.

Feminist methodologies

The authors' different paths to knowledge of feminism and the literature also affected our approach to research design. Sue's stronger grounding meant that she was often more aware of potentially gendered power differentials in focus groups, how these could be addressed, and the importance of empathy with participants in recognition that a universal, 'rational' and scientifically neutral approach was a fallacy born out of a deeply patriarchal research culture. This questioning of a universal truth arose from her understanding of post-structural feminism and highlights the linking of theory and methodology – an important aspect of feminist research.

Just as feminist theory is broad and interdisciplinary, the wide array of feminist methodologies generally seeks to include women's concerns and experiences into research. As feminist researchers, we are mindful to recognise that there are inherent power imbalances between researcher and research participant, and we sought through a feminist research praxis to reduce these imbalances. Just as feminists aim to redress inequities and societal hierarchies, we recognised that researching involves reducing the hierarchy between researcher and participant and we worked to avoid replicating the power dynamics that operate in a traditional masculinist society (Alvesson & Skóldberg, 2018; McHugh, 2014). We were also aware that as feminist researchers, we were to not only recognise the power imbalances between researcher and participant, but also the differences between women, whereby researchers operate from a privileged position due to status, class, race and other identity markers (Gorelick, 2011).

In terms of epistemology, all points of view are represented within feminism, from neo-positivism to postmodernism/poststructuralism (Alvesson & Skóldberg, 2018: 239). As feminist researchers we wanted to challenge traditional concepts of truth embedded within positivism, recognising that knowledge is always impartial and it is not possible for researchers to be impartial or neutral (Gorelick, 2011; McHugh, 2014). Stances of neutrality and impartiality uphold the status quo. Therefore, it was also important for us as researchers adopting a feminist paradigm to not only recognise power differentials within the research process, but to also recognise that we all hold biases, both implicit and explicit (McHugh, 2014). We understood that just as feminists have criticised male researchers' interpretations of women's experiences and the objectification of female participants, feminist researchers are careful to not impose their worldviews and ideologies on participants' responses and experiences (Gorelick, 2011; McHugh, 2014).

Application of the Paradigm

Our focus is the study of women at work. Our recent projects have studied gender equality within Australian public services. We have combined theories of liberal feminism and 'doing gender' in organisations, which have some cross-over. While the feminist theories outlined above identified different types of feminism, the 'doing gender' theories focus on how women, men and organisations create and reinforce gender roles (Abrahamsson, 2014; West & Zimmerman, 1987; Williamson & Colley, 2018). We found these approaches most relevant to our research because they resonated with our personal and professional experiences and our values. Acker's (2006, 2009) work was particularly valuable to us because it allowed us to start from a viewpoint that inequalities occur and accumulate throughout the lifetime – such as a limited initial education affecting career prospects at all future points, or the gender pay gap affecting earnings, independence and retirement prospects – and are present in all policies, processes and meanings. This is more useful than focusing on a single inequality at a particular point in time.

We have used the work of Ely and Meyerson (2000) to frame our study (Williamson & Colley, 2018), as it was important to us to have a framework that captured gender equality in a way that mirrored our own thinking on how we need to dig deeper into structural issues and causes for inequality. Ely and Meyerson (2000) critique the usual approaches to examining gender equity in organisations, based within a liberal feminist framework seeking to pursue gender equity within current institutional structures. They critique the approach of *fix the women*, assuming that women would prosper with the right tools and becoming more like men (Ely & Meyerson, 2000: 105–107). They critique the approach of *value the women*, celebrating their special feminine attributes using a diversity management framework that largely ignores structural barriers to equality and potentially reinforces sex stereotypes (Ely & Meyerson, 2000: 109) and remakes and reinforces the gendering of individuals, jobs and work processes in ways that subordinate women and privilege male traits (Acker, 1990; West & Zimmerman, 1987). They also critique a third approach of creation of policy interventions, such as flexible work arrangements, which may enhance women's working lives but also reinforce stereotypes and domestic responsibilities at work and at home. We welcomed Ely and Meyerson's (2000: 112) critique of these approaches and agreed with them in their summary of the approaches as trying to 'change structures that produce inequality without corresponding interventions into beliefs that legitimate the inequality'. It was refreshing to find that they advocate a framework that disrupts gender through systematic workplace interventions, where workers critique gendered organisational practices, and experiment with new ways of working.

We found their analytical framework suited our intent as researchers in feminism. The framework focuses on four social practices within organisations, which they argue perpetuate gendered social and organisational arrangements (Ely & Meyerson, 2000: 114). The first practice they identify is *formal policies and procedures*, such as legislation, rules and policies. Identifying this practice as gendered resonated with us, as we examined human resource policies and other regulation governing workplaces. Ely and Myerson's second practice is *informal work practices*, norms and patterns of work, which is also highly relevant to our research as we examined who undertakes the different types of work in organisations, and how women and men access and use different working arrangements, such as flexible working arrangements. Their third practice is *narratives, rhetoric and expression*, which is about how workplace culture is based on a model of a male ideal worker. Workplace norms are unstated, but are present in culture, stories and how people contribute to organisational culture. The fourth practice is *informal patterns of everyday social interaction*. This refers to how people relate to each other in organisations, and how people make sense of the world. Often workers do not see the gender dynamics at play, but they are evident in the way people relate to each other, especially in meetings.

With all of these aspects, Ely and Meyerson's framework gave us the tools to explore women at work in a scholarly manner. We chose this framework as it seemed to explain what was occurring in the workplaces we were researching. Applying a theory to practical application requires that you think laterally and play with ideas until you think you have a match between theory, and what you think you might have found in a workplace, or what you think you will find (without making any assumptions).

As can be seen in the report from our larger study of four public services (Williamson *et al.*, 2018), we followed this framework. First, we focused on central and agency level policy frameworks on gender equality, to explore the macro level structures. Then we interviewed both executives and human resource specialists regarding these policy frameworks, to extend our understanding of policy intentions and potential gender blindness in policies. Then we conducted 40 focus groups with 297 managers across four jurisdictions, to test their knowledge on the ground of these policies, their experience in making formal decisions under these policies, their informal practices and patterns of work, and their rhetoric and expressions around gender equity. This approach allowed us to delve deeper into how they think about, talk about and enact policies through their everyday decisions and interactions. Throughout our research project, from designing the project to analysing results and compiling findings, it was also important to ensure that linkages with current feminist theory were maintained. For example, we needed to consider the use of non-binary language and how to recognise the experiences of gender queer participants.

Reflection on the Use of the Paradigm

The feminist paradigm is contested in various ways, from outright resistance to subtle reframing into less effective approaches such as neoliberal feminism. Therefore, using the paradigm can be fraught, and even the most ardent feminists (like us) will tread very carefully. This includes the way we frame the research project to our research partners, identifying ways to appeal to their interest in gender equity without seeming ready to stage a revolution. It also means framing the research methods and focus group form so as not to get participants offside or afraid to speak about their concerns, doubts or confusions around gender equity.

We favour qualitative research such as focus groups, to provide the opportunity for in-depth exploration of ideas and to allow us to also pursue additional lines of inquiry as warranted. Enabling employees to participate in focus groups incorporates a feminist approach which enables women's voices to be heard, for participants to share and discuss experiences and even result in a form of 'consciousness raising' (Gorelick, 2011; Smithson, 2009). This approach minimises the role of the researcher (McHugh, 2014), limits subjectivity and interpretation and thereby can further reduce power differences.

As we study gender equality in large organisations, this means extensive fieldwork. For example, in our recent project we had four public services as research partners (Queensland, New South Wales, South Australia and Tasmania), and studied two agencies in each of those jurisdictions, and aimed for 40 participants from each agency. While time-consuming, expensive in travel terms and emotionally exhausting, it fielded rich data that have supported the development of useful reports for each jurisdiction, an over-arching report, and a lot of potential research publications.

Before commencing fieldwork, we considered the assumptions we held about the gender equity in the case study organisations and carefully examined our questions and approaches to limit biases. For example, we considered our assumptions about the organisations under study and the people who worked there. We were careful in focus groups to not impose our feminist values. This was challenging at times, especially in groups conducted in male-dominated agencies, where concepts of gender equality were regularly questioned, or we were told the gender issues had 'been fixed already'. Further, we adopted a feminist approach by encouraging women within the groups to speak openly and to explore the questions in their own way. Providing women with this space enabled them to challenge some of the anti-feminist statements made by their male colleagues, thereby empowering the women. Our research therefore not only contributes to a feminist body of knowledge, but also assists women in expressing their voice – an important aim of feminist ideologies.

Conclusion

So, what is our message to new researchers on how feminist research be conducted? Here is some general advice and principles from our accumulated knowledge and experience:

- Include gender analysis that makes women visible.
- Let women speak for themselves, recognising the importance of voice.
- Reduce power differentials between yourself as a researcher and the participant. For example, exchange stories about the topic under investigation, which can make the relationship more reciprocal.
- Test why progress has stalled under new models such as gender mainstreaming.
- Consider inequality regimes as an analytic approach.
- Avoid positivist approaches or quantitative approaches that ignore the power dimensions and assume or generalise as if there is a single truth for all women.
- Treat gender as a verb rather than a noun, such as gendering of processes or organisations.

When designing your research project, make sure you are not overly reliant on previous research and literature that is male dominated. If you have very few references to female authors, you need to consider whether your project is gender biased (Alvesson & Skóldberg, 2018).

And when people criticise you for studying women in isolation, do not be afraid to argue that it is legitimate. Too often, women are studied and compared to men, as if the male experience is the valid starting point against which women should be measured. It is okay to just study women's experiences but be ready to explain and defend that. Also consider how to get the best information. If you want women to talk about their experience of discrimination, it is more likely they will be forthcoming without men in the group, especially in a work setting where there are career as well as power considerations.

Our next challenge is to make sense of a newer phenomenon that is being broadly labelled gender fatigue (but includes other terms such as gender blindness, gender suppression and gender resistance and backlash), which describe organisational weariness of trying to address gender inequality over so many years, as well as some resistance from men that things have swung too far in favour of women. This literature is new, and we are trying to assist to make sense of these various terms. Our feminist journey continues.

References

Abrahamsson, L. (2014) Gender and the modern organization, ten years after. *Nordic Journal of Working Life Studies* 4 (4), 109–136.

Acker, J. (1990) Hierarchies, jobs, bodies: A theory of gendered organizations. *Gender and Society* 4 (2), 139–158.

Acker, J. (2006) Inequality regimes: Gender, class, and race in organizations. *Gender and Society* 20 (4), 441–464.

Acker, J. (2009) From glass ceiling to inequality regimes. *Sociologie du Travail* 51 (2), 199–217.

Alvesson, M. and Skóldberg, K. (2018) *Reflexive Methodology: New Vistas for Qualitative Research* (3rd edn). Los Angeles: Sage.

Calvini-Lefebvre, M., Cleall, E., Grey, D.R., Grainger, A., Hetherington, N. and Schwartz, L. (2010) Rethinking the history of feminism. *Women: A Cultural Review* 21 (3), 247–250. doi:10.1080/09574042.2010.516906

Cho, S., Crenshaw, K.W. and McCall, L. (2013) Toward a field of intersectionality studies: Theory, applications, and praxis. *Signs* 38 (4), 785–810.

Church Gibson, P. (2004) Introduction: Popular culture. In S. Gillis, G. Howie and R. Munford (eds) *Third Wave Feminism: A Critical Exploration*. Basingstoke: Palgrave Macmillan.

Colley, L. and White, C. (2019) Neoliberal feminism: The neo-liberal rhetoric about feminism by Australian political actors. *Gender Work and Organization* 26 (8), 1083–1099.

Curthoys, A. (1988) *For and Against Feminism: A Personal Journey into Feminist Theory and History*. Sydney: Allen & Unwin.

Denis, A. (2008) Review essay: Intersectional analysis: A contribution of feminism to sociology. *International Sociology* 23 (5), 677–694.

Ely, R.J. and Meyerson, D.E. (2000) Theories of gender in organizations: A new approach to organizational analysis and change. *Research in Organizational Behavior* 22, 103–151.

Faludi, S. (1991) *Excerpt from Backlash: The undeclared war against American women,* http://susanfaludi.com/backlash.html: Crown.

Ford, C. (18 June 2018) The 'not all men' excuse is absurd. *The Sydney Morning Herald,* https://www.smh.com.au/lifestyle/life-and-relationships/the-not-all-men-excuse-is-absurd-20180618-p4zm94.html.

Fraser, N. (2009) Feminism, capitalism and the cunning of history. *New Left Review* 56, 97–117.

Gorelick, S. (2011) Contradictions of feminist methodology. In P. Atkinson and S. Delamont (eds) *SAGE Qualitative Research Methods* (pp. 343–360). Los Angeles: Sage.

Hawkesworth, M. and Disch, L. (2016) Feminist theory: Transforming the known world. In M. Hawkesworth and L. Disch (eds) *The Oxford Handbook of Feminist Theory* (pp. 1–15). Oxford: Oxford University Press.

Hemmings, C. (2005) Telling feminist stories. *Feminist Theory* 6 (2), 115–139.

Kirton, G. and Greene, A (2015) *The Dynamics of Managing Diversity.* London: Routledge.

McHugh, M. (2014) Feminist qualitative research: Toward transformation of science and society. In P. Leavy (ed.) *The Oxford Handbook of Qualitative Research.* Oxford: Oxford University Press.

McKnight, L. (2016) A bit of a dirty word: 'Feminism' and female teachers identifying as feminist. *Journal of Gender Studies* 1–11. doi:10.1080/09589236.2016.1202816

McRobbie, A. (2009) *The Aftermath of Feminism: Gender, Culture, and Social Change.* London: Sage.

Moses, C. (2012) 'What's in a Name?' On writing the history of feminism. *Feminist Studies* 38 (3), 757–779.

Noon, M. (2007) The fatal flaws of diversity and the business case for ethnic minorities. *Work, Employment and Society* 21 (4), 773–784.

Roiphe, K. (1993) *The Morning After: Sex, Fear and Feminism.* London: Hamish Hamilton.

Rottenberg, C. (2014) *The Rise of Neoliberal Feminism.* Abingdon: Routledge.

Sandberg, C. and Scovell, N. (2015) *Lean-In: Women, Work and the Will to Lead.* London: WH Allen.

Sawer, M. (2014) Gender equality architecture: The intergovernmental level in federal systems. *Australian Journal of Public Administration* 73 (3), 361–372.

Schein, R. (2014) Hegemony not co-optation: For a usable history of feminism. *Studies in Political Econom,* 94 (1), 169–176. doi:10.1080/19187033.2014.11674959

Smithson, J. (2009) Focus groups. In P. Alasuutari, L. Bickman and J. Brannen (eds) *The SAGE Handbook of Social Research Methods* (pp. 375–370). Los Angeles: Sage.

Strachan, G., Burgess, J. and Henderson, L. (2007) Equal employment opportunity legislation and policies: The Australian experience. *Equal Opportunities International* 26 (6), 525–540.

UN (2018) United Nations Sustainable Development Goals, accessed at https://www.un.org/sustainabledevelopment/sustainable-development-goals/.

Walby, S. (2005) Gender mainstreaming: Productive tensions in theory and practice. *Social Politics: International Studies in Gender, State and Society* 12 (3), 321–343.

West, C. and Zimmerman, D.H. (1987) Doing gender. *Gender and Societ* 1 (2), 125–151.

Williamson, S. and Colley, L. (2018) Gender in the Australian public service: Doing, undoing, redoing or done? *Australian Journal of Public Administration* 77 (4), 583–596.

Williamson, S., Colley, L., Foley, M. and Cooper, R. (2018) *The Role of Middle Managers in Progressing Gender Equity in the Public Sector.* Final Report for Australian New Zealand School of Government project, June 2018.

11 The Pragmatic Paradigm in Destination Competitiveness Studies: The Case of the SCUBA Diving Tourism Niche

Ambrozio Queiroz Neto, Gui Lohmann, Noel Scott and Kay Dimmock

Introduction

In this chapter, we discuss how the pragmatic paradigm shaped an investigation of destination competitiveness and customer value in a tourism niche market. We describe how this paradigm guided the implementation of a mixed-method approach consisting of two sequential stages: first semi-structured interviews and then an online survey. The interviews provided a view of SCUBA diving tourists' thoughts on what makes a successful SCUBA diving destination. The quantitative survey evaluated differences in the perceptions of different groups of SCUBA diving tourists about what makes a successful SCUBA diving destination.

Review of the Pragmatic Paradigm

A pragmatic paradigm seeks to enhance understanding of the complex social world (Pansiri, 2006), by fostering a dialogue between philosophical/theoretical and empirical incongruences, particularly in mixed methods research. As Morgan (2014: 7) notes, 'for most of the researchers operating within the field of mixed-method research, the appeal of pragmatism…[is] about its practicality'. The word pragmatism derives from *pragma* (Greek) and means *practice* and *practical*. Ontologically, the pragmatic paradigm is classified as 'what works' in the empirical world (Jennings, 2010). As summarised by Khoo-Lattimore *et al.* (2019: 1537), in the pragmatic paradigm, 'research designs and strategies should privilege the most effective ways of answering research questions over philosophical assumptions'.

In his article 'Competitive advantage: logical and philosophical considerations', Powell (2001: 884) argues that

> [t]o a pragmatist, the mandate of science is not to find truth or reality, the existence of which are perpetually in dispute, but to facilitate human problem-solving. According to pragmatist philosopher John Dewey, science should overthrow the notion, which has ruled philosophy since the time of the Greeks, that the office of knowledge is to uncover the antecedently real, rather than, as is the case with our practical judgments, to gain the kind of understanding which is necessary to deal with problems as they arise.

From an epistemological perspective, pragmatism refutes the idea that 'truth' can be determined once and for all. 'Truth is what works', and knowledge cannot be entirely abstract from contingent beliefs, interests and projections (Howe, 1988). Hence, for the researcher adopting the paradigm of pragmatism, there is a quest to ontologically determine and understand what the 'truth' is or can be known as and equally, to realise that how we find the 'truth' or knowledge is through experiences and actions, including that of the researcher.

The methodological framework frequently adopted in pragmatic paradigm projects comprises the two extremes of positivism/post-positivism and interpretivism: quantitative methods are utilised by the former and qualitative methods by the latter. As a result of this combination, the pragmatists are considered the founders of the mixed-methods approach to data collection (Tashakkori & Teddlie, 1998, 2003). Kaushik and Walsh (2019: 2) point out that pragmatism is able to close the gap between the scientific method approach (a quantitative one that fosters objectivity and deductive reasoning), with 'structuralist orientation of older approaches', and the naturalistic methods (a qualitative one that targets participatory action, and rhetoric of advocacy and change), characterised by 'freewheeling orientation of newer approaches'. As the first author, Ambrozio, became familiar with pragmatism, he came to realise this was the paradigm that would be more aligned with his research on destination competitiveness (DC). Examination of studies relating to the concept of DC, it further strengthened the need for this research to engage with pragmatism.

Pragmatism is common among the research paradigms used in tourism studies, particularly those that are supported by mixed or multiple methods (Khoo-Lattimore *et al.*, 2019; Pansiri, 2006). Some of the examples of mixed methods approaches in tourism studies include:

- Mixed-methods of systematic reviews, bringing together both quantitative and qualitative meta-analysis. Meta-analysis, as the name suggests, involves the analysis of analyses. They have been extensively used in tourism due to the current stage of knowledge consolidation that a still young and growing discipline continues to face. Either through narrative or systematic literature reviews, this method involves a combined quantitative and qualitative approach (Gretzel & Kennedy-Eden, 2012). Examples of mixed-methods systematic reviews include the work done by Spasojevic *et al.* (2018) analysing 157 journal articles on air transport and tourism, and Le *et al.* (2019) provided a theoretically based review of imagery processing research in tourism, reviewing and analysing 70 relevant papers published in the period from 1997 to 2017.

- Network analysis and evaluation methods aim to examine the relationships of actors that operate and collaborate in a network, in particular the impact that (1) actors have on each other, (2) the individual actors have on a network, (3) the network has on individual actors and (4) the interactions of the various actors have on the whole network (Ahmed, 2012). These network arrangements are very typical of the tourism literature, both in terms of the physical connections that facilitate travel to occur, as well as the communication and business channels used to transmit data/information.
- Supply and demand mixed methods: the multi-stakeholder nature of the tourism industry usually requires the understanding of a certain phenomenon through the lenses of several actors. Research on tourists' choices, behaviour and impacts is commonly based on statistical and quantitative analysis (Leung *et al.*, 2017). There are also a substantial number of studies that undertake an interpretivism approach, usually describing a given market (Mazanec, 1984), management issues and the strategies used by tourism providers (Lohmann & Pearce, 2012).

This chapter builds on this past work in tourism and highlights how pragmatism can inform methodological choices in research design, especially in the implementation of a mixed-methods approach.

Pragmatism in the Context of a PhD Research on Destination Competitiveness

Ambrozio's study examined the concept of DC from an alternative perspective; one that considers DC not only in terms of the competitiveness indicators assessing the suppliers' side of a destination, but also one that adds value to the customer. More specifically, his PhD investigated customer value in DC in the context of SCUBA diving tourism, a niche tourism market. He used mixed-methods, deploying face-to-face interviews and online surveys.

The concept of DC was developed in the 1990s (Crouch & Ritchie, 1993, 1999; Kozak & Rimmington, 1999; Poon, 1993;) to satisfy the need for new management tools to deal with an economic crisis of the period (Poon, 1993). The diamond model for competitive advantage (Porter,

1990) and the resource-based view of the firm (Penrose, 1959; Wernerfelt, 1995) from management are the basis of the development of tourism DC models. The variation in models suggested that the paradigm relating to this research would need to be able to capture the different perspectives.

The abovementioned authors highlight that competitiveness leads to a sustained improvement in the well-being of residents, represented by economic prosperity, environmental stewardship and resident quality of life. D'Hauteserre (2000: 23) defines DC as 'the ability of a destination to maintain its market position and share and/or improve upon them through time'. Moreover, D'Hauteserre (2000) extends the concept of competitiveness to product life cycle, arguing that a company that seeks to be more competitive will extend, at the same time, their product lifetime, postponing a decline. Hassan (2000) argues that DC is the ability to create and integrate value-added products that sustain resources while maintaining market position relative to competitors.

These different perspectives on DC result in different models to frame DC (Table 11.1). Crouch and Ritchie (1993, 1999) and Crouch (2011) are widely reported and used as a basis for other relevant studies in the field. Queiroz Neto *et al.* (2017) summarise Crouch and Ritchie's seminal work in three phases: (1) conception of ideas (Crouch & Ritchie, 1993), (2) the consolidation of a conceptual model (Crouch & Ritchie, 1999) and (3) ranking of determinant attributes used in their model (Crouch, 2011). Influenced by the Crouch and Ritchie (1999) model, Dwyer and Kim (2003) present a DC model recognising demand as an essential element (previously unobserved). Mostly, DC models are developed based on a suppliers' perspective.

Wilde (2010), Pabel and Coghlan (2011) and Andrades-Caldito *et al.* (2014) investigated the importance placed on DC attributes from the demand side. Initially, Wilde (2010) examined the importance given to 38 destination attributes by tourists that visited two destinations in Australia: the Coffs Harbour Coast and Great Lakes. Pabel and Coghlan (2011) measured the importance and performance of 16 dive trip attributes from the perspective of divers visiting the Great Barrier Reef in Cairns, Australia. In this case, attributes such as accommodation, restaurants and infrastructure were not considered in the study. Lastly, Andrades-Caldito *et al.* (2014) analysed the importance of 13 destination attributes from the view of tourists who visited the region of Andalucía, Spain.

Focusing on the suppliers' side, Hong (2009) and Crouch (2011) also investigated the importance of DC attributes. Both authors use their model of DC. Table 11.1 shows the comparison of the results obtained by above-mentioned studies. Due to different terminologies used in the reviewed studies, the comparison is limited to groups of attributes.

Abreu-Novais *et al.* (2016) highlighted that the complexity of investigating DC results in knowledge gaps, especially from the customer's perspective. Many models are fundamentally based on a suppliers' perspective

Table 11.1 Analysis of important destination competitiveness attributes

Author/Reference	Based on	Perspective	Most important factors
Hong (2009)	Hong (2009)	Suppliers/Experts 19 respondents	1. Exogenous comparative advantages 2. Endogenous comparative advantages 3. Competitive advantage 4. Tourism management 5. Domestic environment conditions 6. Global environment conditions
Wilde (2010)	Dwyer and Kim (2003)	Demand 344 respondents	1. Destination management (public and private) 2. Facilities and activities 3. Nature 4. Augmented benefits 5. History
Crouch (2011)	Crouch and Ritchie (1999)	Suppliers/Experts 83 respondents	1. Core Resources and attractors 2. Destination Management 3. Qualifying and amplifying determinants 4. Destination policy planning and development 5. Supporting factors and resources
Pabel and Coghlan (2011)	Dwyer and Kim (2003)	Demand 296 respondents	1. Environmental attributes 2. Setting attributes 3. Service attributes
Andrades-Caldito et al. (2014)	Crouch and Ritchie (1999); Dwyer and Kim (2003)	Demand 4195 respondents	1. Created resources 2. Endowed resources 3. Tourists' perceptions about destination management

Source: Queiroz Neto et al. (2017).

(Figure 11.1). Woodruff (1997) emphasised that customer value is an important source of competitive advantage and, rather than being delivered by suppliers, value is co-created by the customer (Vargo & Lusch, 2004). An investigation of the destination attributes valued by tourists is a relevant gap (Queiroz Neto et al., 2016) and was the main aim of Ambrozio's PhD research.

Reviewing previous works on DC (Figure 11.1), two different phases were identified: an investigation of destination attributes/developing models on DC and measurement of the importance of DC/ranking destination attributes. These two phases implement two sequential research methods: qualitative followed by a quantitative approach. Firstly, while Crouch and Ritchie (1993, 1999) used a series of interviews and workshops, Dwyer and Kim (2003) used a literature review to develop their DC model. In the second phase, different quantitative methods were implemented to assess the importance of DC attributes: Wilde (2010) and Pabel

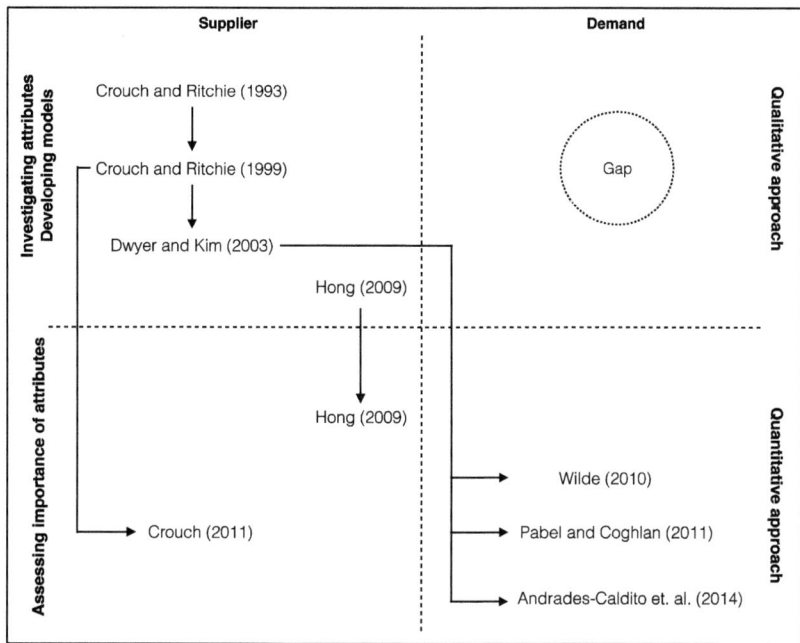

Figure 11.1 Identification on knowledge gap

and Coghlan (2011) used a survey; Crouch (2011) implemented an analytical hierarchy process; and Andrades-Caldito *et al.* (2014) used structural equation modelling. Hong (2009) consolidated two stages in the same study: a literature review to establish his model of DC (qualitative); and an analytical hierarchy process to measure the importance of the destination attributes of his model (quantitative).

The customer value determination process (Woodruff, 1997) is composed of five guiding questions: (1) what do customers value?; (2) of all the value dimensions that targeted customers want, which are the most important?; (3) how well or poorly is the supplier doing in delivering the value that targeted customers want? (4) why are suppliers doing poorly (well) on important value dimensions?; and (5) what are the targeted customers likely to value in the future? These five guiding questions can be implemented in three sequential stages: investigation of attributes (question 1), measurement of importance of attributes (questions 2 and 5) and measurement performance (questions 3 and 4).

Disregarding the measurement of a destination's performance, both DC studies and customer value determination processes follow similar approaches: a qualitative stage to investigate (destination) attributes; and a quantitative stage to measure the importance of (destination) attributes. Hence, the implementation of a research paradigm that could embrace these two sequential stages was critical to guide the investigation of DC through customer value. Therefore, pragmatism was presented as the most suitable paradigm to guide and shape this study.

Research design

The research therefore aimed to investigate DC from a customer value perspective. Indeed, it considered that there are different customer value perspectives related to the levels of travel experience and experience in a leisure activity (in this case, diving experience). Therefore, this study examined, in particular, the customer perspective of value in a niche market: scuba diving tourism. To achieve this aim, Ambrozio's PhD research brought together two broad theories: DC and customer value. In support of this aim, four research questions were proposed:

RQ1: What are the destination attributes scuba diving tourists value in a successful scuba diving destination?
RQ2: From the perspective of scuba diving tourists, what is the relative importance of each DC attribute?
RQ3: How do scuba diving tourists with different travel experience levels differ in the importance given to DC attributes?
RQ4: How do scuba diving tourists with different diving experience levels differ in the importance given to DC attributes?

In line with the paradigm of pragmatism, this exploratory research implemented a mixed-method approach in two sequential stages: qualitative and quantitative. In the first stage, to answer research question one (RQ1), semi-structured interviews were applied. Afterward, in the second stage, an online survey was implemented to answer research questions 2, 3 and 4 (RQ2, RQ3 and RQ4). Figure 11.2 showcases the conceptual framework designed specifically for this research.

First stage: interviewing SCUBA divers

Influenced by Woodruff's (1997) customer value determination approach and Crouch and Ritchie (1999), the first stage aimed to answer RQ1. The idea of 'successful destination' was first applied by Crouch and Ritchie (1999) while developing their DC model.

Woodruff and Gardial (1996) indicate two different techniques to investigate customer value: focus groups or interviews. Initially, this research aimed to use focus group interviews. However, during the pilot study, two issues were perceived with the proposed approach. Firstly, as pointed out by Gabott and Hogg (1998), value is individual and varies according to very particular reasons of a person. Thus, a thorough investigation was not possible during focus groups due to the complexity of value. Secondly, SCUBA diving tourists, in general, want to spend most of their holidays relaxing and resting, when not diving. Therefore, gathering

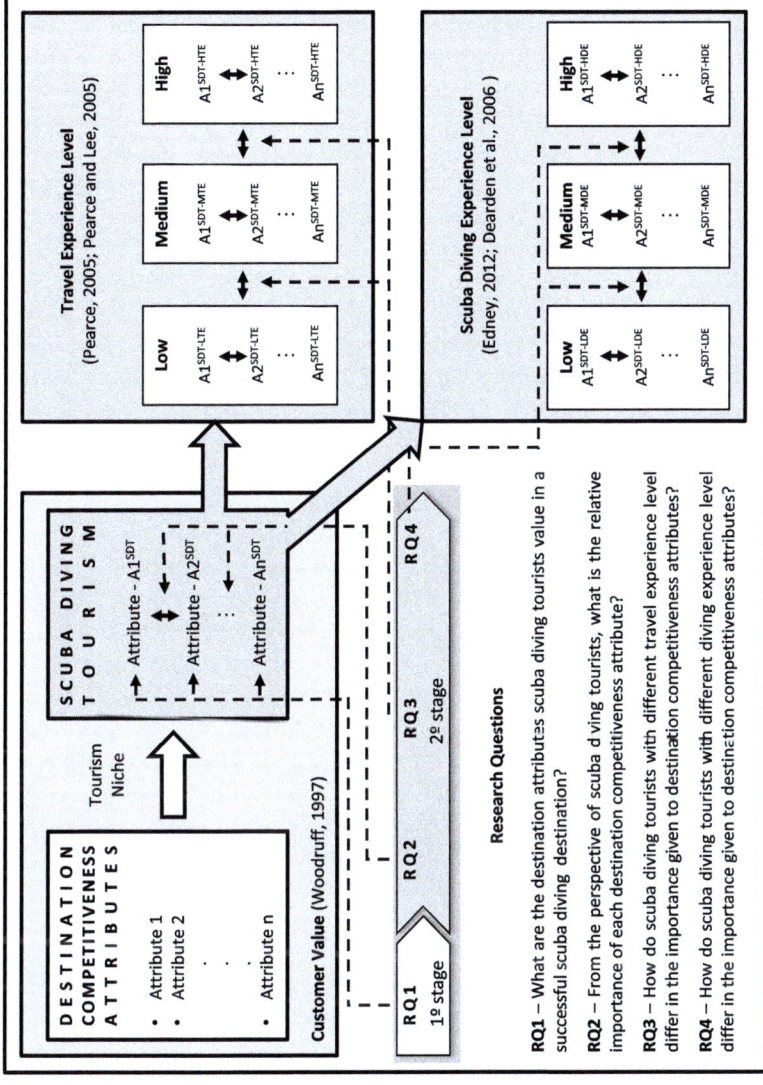

Figure 11.2 Research framework

a particular group of interviewees at the same time was challenging with tourists during their holidays. For these reasons, after analysing the results obtained during the pilot study, the researchers decided to swap the method from focus groups to semi-structured interviews. Further, semi-structured interviews were recommended when there is no chance to interview the respondent in a future moment (Bernard, 2000). Hence, it was the most suitable interviewing technique to apply to tourists during their holidays.

Regarding the sampling approach, a non-random convenience sample (Gideon, 2012) was applied. The reason for the choice was that a convenience sample was a useful method for exploratory studies (Neuman, 2011) and it involves seeking out individuals meeting specific criteria (Gideon, 2012). In the case of this study, certified SCUBA divers that travelled to tourism destination to undertake SCUBA diving activities were designated.

To answer RQ1 and to be in line with a pragmatic paradigm, two distinct approaches were implemented: deductive and inductive. First, SCUBA diving tourists were asked to 'think about two of the most successful SCUBA diving destinations' they have visited in the past. After addressing two destinations, respondents were asked to list the attributes that made those past experiences successful. Second, interviewees were asked to 'think about the most successful (hypothetic) SCUBA diving destination'. Afterwards, a handout with a list of destination attributes based on previous studies (Dearden *et al.*, 2006; Hong, 2009; Pabel & Coghlan, 2011) was presented to the respondents. The respondents could either confirm, delete or add new attributes. In the end, divers were asked to describe the reasons, importance and unimportance of the listed destination attributes. Overall, the instrument of data collection was composed of six enquiries related to (1) diving experience; (2) travel experience to SCUBA diving destinations; (3) influence of diving activity in travel decision making; (4) destination attributes of successful SCUBA diving destinations visited in the past; (5) destination attributes for the most successful SCUBA diving destination (hypothetic); (6) reasons, importance and unimportance of listed destination attributes. The application of these two approaches undoubtedly provided a comprehensive picture of the attributes of a successful SCUBA diving destination.

Data collection for the first stage occurred from August to October of 2015 in four popular SCUBA diving destinations: Gold Coast and The Great Barrier Reef (Cairns) in Australia and Phuket and Koh Tao in Thailand. In total, 34 SCUBA diving tourists were interviewed during the period of data collection. The researchers targeted SCUBA divers with different levels of diving experience and a balanced number per gender to get a comprehensive list of destination attributes. The first author was encouraged to find that the profile of interviewees (Table 11.2) was similar to the profile of respondents of previous studies (Dearden *et al.*, 2006; Garrod, 2008; Pabel & Coghlan, 2011).

Table 11.2 Profile of respondents in stage one

Level of diving experience	Gender		Total per diving experience	(%) per diving experience
	Male	Female		
Low diving experience	9	6	15	(44%)
Medium diving experience	5	3	8	(24%)
High diving experience	8	3	11	(32%)
Total per gender	22	12	34	–
(%) per gender	(65%)	(35%)	–	(100%)

To answer RQ1, thematic analysis was implemented. Thematic analysis is a method for identifying, analysing and reporting patterns (themes) within data. Fereday and Muir-Cochrane (2006: 82) argue that thematic analysis is a 'form of recognition within the data where emerging themes become the categories for analysis'. With the assistance of NVivo 11, the six steps of thematic analysis (Braun & Clarke, 2006) were implemented: (1) familiarising with the data; (2) generating initial codes; (3) searching for themes; (4) reviewing themes; (5) defining and naming themes; (6) producing final report.

The results obtained in the first stage of the study fed into the design of the questionnaire that was applied in the second stage. The details of methodological steps and results of the first stage were published in Queiroz Neto *et al.* (2018).

Second stage: online survey

SCUBA divers are well educated (Dearden *et al.*, 2006; Edney, 2012); a significant number of them live in Europe and the United States of America (Garrod, 2008). SCUBA divers form a large global community. Evidence of this can be seen on social networking sites such as Facebook. Facebook groups relating to SCUBA diving have hundreds of thousands of members. Because of the ability to reach members of this community through social media, the second stage of this study involved the administration of a questionnaire-based online survey to answer RQ2, RQ3 and RQ4. An online questionnaire survey containing closed and open-ended questions was provided to certified SCUBA divers. Data collection took place from January to June of 2016.

Online surveys are popular as a research method nowadays. However, there are several strengths and potential weaknesses in the implementation of online surveys (Evans & Mathur, 2005). While global reach, flexibility, convenience and the opportunity to reach controlled samples are considered strengths; sample selection, impersonal approach, lack of online experience from the respondents and low response rate can be potential weaknesses for online surveys.

As in the first stage, certified SCUBA divers (SCUBA diving tourists) were the target population for data collection. In seeking a significant sample size to provide a high degree of accuracy, the rationale for the sampling size was based on the formula developed by Krejcie and Morgan (1970) and other quantitative approaches presented in similar studies. While the method of Krejcie and Morgan (1970) indicates 387 sample units are required (for an unknown population), similar studies use different sample units: Kim *et al.* (2008) when segmenting golf tourists used 370 respondents, and Sung (2004) examining the consumer and travel behaviour of adventure travellers used 892 sample units. In the end, the total number of respondents in this study was 712 sample units.

This research used variety of strategies to recruit SCUBA divers for the online survey: (1) a website (www.scubabestchoice.com) was developed to publicise the details of the research: research aims and objectives, profile of researchers, contact and link to the questionnaire; (2) a banner (in four languages) and a 40 second video were developed and posted on Facebook and Twitter; (3) two paid advertisement campaigns on Facebook framing individuals who liked 'SCUBA diving activities' in their profile. The implementation of these different and simultaneous strategies was fundamental to the achievement of the planned number of respondents.

The survey instrument was designed based on analysis of data from the previous stage of this study and on previous studies (Dearden *et al.*, 2006; Edney, 2012; Pearce, 2005; Pearce & Lee, 2005). The questionnaire comprised five sections: socio-demographics, travel experience, SCUBA diving experience, SCUBA diving trip characteristics and assessment of DC attributes. For the section 'assessment of DC attributes', the 89 destination attributes obtained in the first stage were reduced to 52 attributes. A pilot test ($n = 20$) was performed in November of 2016 and as a result some questions were re-worded or reformulated. The final questionnaire included 19 questions.

In this stage, analytical procedures for quantitative data varied according to the research questions. Firstly, for research RQ2, descriptive analysis and principal components analysis were implemented. Secondly, for RQ3 computation of travel experience score, descriptive analysis, k-means cluster analysis, cross-tabulation and analysis of variance (ANOVA) with post-hoc pairwise comparisons were applied. Finally, descriptive analysis, cross tabulation and analysis of variance with post-hoc pairwise comparison were used to analyse RQ4.

To highlight the structure of the data analysis implemented on this research, an analytical framework was created. Figure 11.3 was designed to showcase step-by-step the methods applied in the study.

The results obtained in the second stage of this study were published in two journal articles. Results for RQ2 and RQ3 are published in

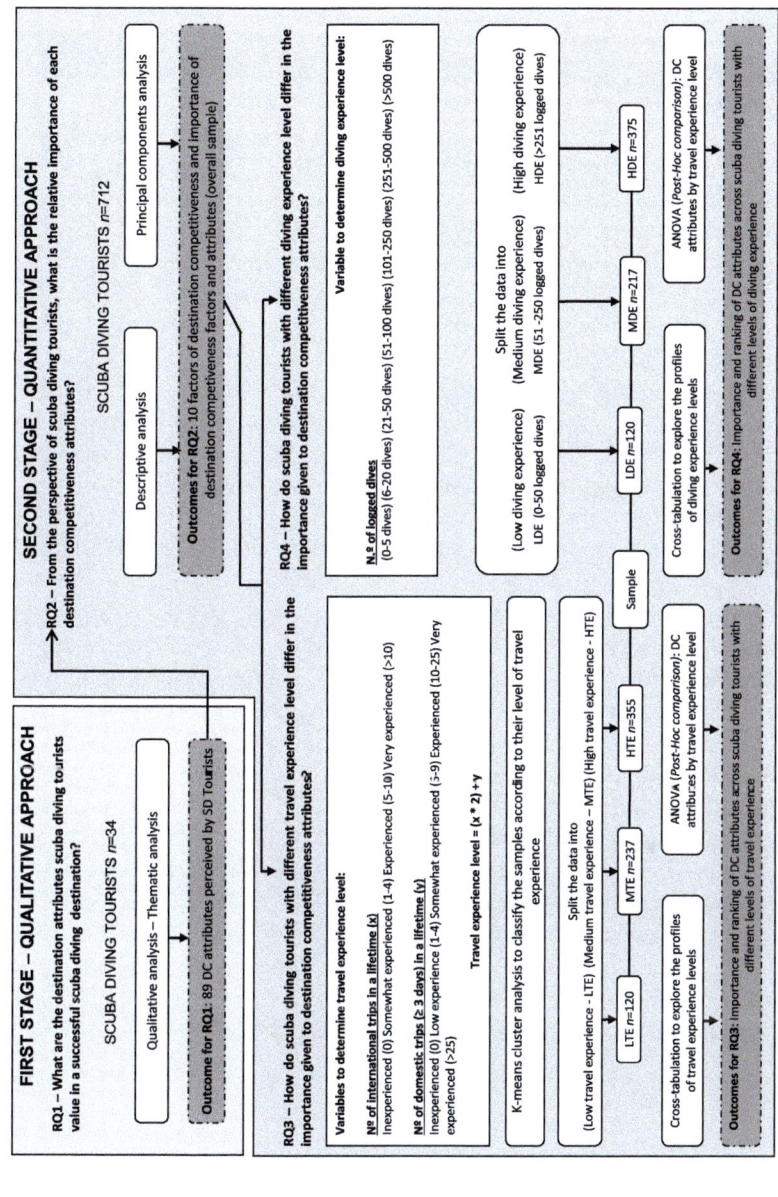

Figure 11.3 Methodological framework

Queiroz Neto *et al.* (2017) and results for RQ4 are presented in Queiroz Neto *et al.* (2018).

Reflection on the Use of the Paradigm

Choosing a paradigm is one of the most important tasks for the researcher while shaping a study. The decision should be based on the research questions proposed (Stockman, 2015). Following this premise, pragmatism was the most suitable paradigm because it covered both qualitative and quantitative approaches (led by the research questions). In the case of this investigation, the application of pragmatism provided (1) a better understanding of how knowledge on the topics (DC, customer value and SCUBA diving tourism) were forged; (2) as a consequence of the first, a clear evaluation on the level of maturity of the studied area; and (3) a guidance for the development of the research design and its procedures.

Every research has predictable and unpredictable issues regardless of the paradigm. The use of pragmatism requires good planning in order to fit within the research's timeframe. The researcher must be aware of the processes while applying a mixed method approach. Usually, one stage leads to the other. It means that the researcher needs discipline to guarantee the rigour in every research step. Otherwise, the research could lack rigour or might end up becoming a never-ending essay.

Through the application of a mixed-method approach with two sequential stages, the first author was able to capture the complexity of perspectives when studying the demand's point-of-view. The qualitative method, applied in the first stage of the study, provided the perspectives of SCUBA diving tourists towards attributes of a successful SCUBA diving destination. The final stage investigated the different perceptions among SCUBA diving tourists towards DC attributes. Howe (1988) writes that knowledge cannot be abstracted from contingent beliefs, interests and projections. Therefore, contemplating the world through the idea that 'truth is what works' brought unique insights to scientific investigation. In this way, a pragmatic paradigm allows a better understanding of the complexity in tourism and hospitality studies. Thus, the authors will keep using pragmatism when the research design requires a more comprehensive approach to investigate what works in reality.

References

Abreu-Novais, M., Ruhanen, L. and Arcodia, C. (2016) Destination competitiveness: what we know, what we know but shouldn't and what we don't know but should. *Current Issues in Tourism* 19 (6), 492–512.

Ahmed, E. (2012) Network analysis. In N. Seetaram, A. Gill and L. Dwyer (2012) *Handbook of Research Methods in Tourism: Quantitative and Qualitative Approaches* (pp. 472–494). Cheltenham: Edward Elgar Publishing

Andrades-Caldito, L., Sánchez-Rivero, M. and Pulido-Fernández, J.I. (2014) Tourism destination competitiveness from a demand point of view: An empirical analysis for Andalusia. *Tourism Analysis* 19 (4), 425–440.
Bernard, H.R. (2000) *Social Research Methods: Qualitative and Quantitative Approaches*. Thousand Oaks: Sage.
Braun, V. and Clarke, V. (2006) Using thematic analysis in psychology. *Qualitative Research in Psychology* 3 (2), 77–101.
Crouch, G.I. (2011) Destination competitiveness: An analysis of determinant attributes. *Journal of Travel Research* 50 (1), 27–45.
Crouch, G.I. and Ritchie, J.R.B. (1993, October, 17–23) Competitiveness in international tourism: A framework for understanding and analysis. Paper presented at the 43rd congress of the association internationale d'experts scientixque du tourisme, San Carlos de Bariloche, Argentina.
Crouch, G.I. and Ritchie, J.R.B. (1999) Tourism, competitiveness, and societal prosperity. *Journal of Business Research* 44, 137–152.
D'Hautesserre, A.-M. (2000) Lessons in managed destination competitiveness: The case of Foxwoods Casino Resort. *Tourism Management* 21 (1), 23–32.
Dearden, P., Bennett, M. and Rollins, R. (2006) Implications for coral reef conservation of diver specialization. *Environmental Conservation* 33 (4), 353–363.
Dwyer, L. and Kim, C. (2003) Destination competitiveness: Determinants and indicators. *Current Issues in Tourism* 6 (5), 369–414.
Dwyer, L., Mellor, R., Livaic, Z., Edwards, D. and Kim, C. (2004) Attributes of destination competitiveness: A factor analysis. *Tourism Analysis* 9 (1), 91–101.
Edney, J. (2012) Diver characteristics, motivations, and attitudes: Chuuk Lagoon. *Tourism in Marine Environments* 8 (1/2), 7–18.
Evans, J.R. and Mathur, A. (2005) The value of online surveys. *Internet Research* 15 (2), 195–219.
Fereday, J. and Muir-Cochrane, E. (2006) Demonstrating rigor using thematic analysis: A hybrid approach of inductive and deductive coding and theme development. *International Journal of Qualitative Methods* 5 (1), 80–92.
Gabott, M. and Hogg, G. (1998) *Consumer and Services*. Chichester: Wiley.
Garrod, B. (2008) Market segments and tourist typologies for diving tourism. In B. Garrod and S. Gössling (eds) *New Frontiers in Marine Tourism: Diving Experiences, Sustainability, Management*. Amsterdam: Elsevier.
Gideon, L. (2012) *Handbook of Survey Methodology for the Social Sciences*. New York: Springer.
Gretzel, U. and Kennedy-Eden, H. (2012) Meta-analyses of tourism research. In N. Seetaram, A. Gill and L. Dwyer (eds) *Handbook of Research Methods in Tourism: Quantitative and Qualitative Approaches* (pp. 459–471). Cheltenham: Edward Elgar Publishing.
Hassan, S.S. (2000) Determinants of market competitiveness in an environmentally sustainable tourism industry. *Journal of Travel Research* 38: 239–245.
Hong, WC. (2009) Global competitiveness measurement for the tourism sector. *Current Issues in Tourism* 12 (2), 105–132.
Howe, K.R. (1988) Against the quantitative-qualitative incompatibility thesis or dogmas die hard. *Educational Researcher* 17 (8), 10–16.
Jennings, G. (2010) *Tourism Research* (2nd edn). Milton: Wiley.
Kaushik, V. and Walsh, C.A. (2019) Pragmatism as a research paradigm and its implications for social work research. *Social Sciences* 8 (9), 1–17.
Kim, S.S., Kim, J.H. and Ritchie, B.W. (2008) Segmenting overseas golf tourists by the concept of specialization. *Journal of Travel and Tourism Marketing* 25 (2), 199–217.
Khoo-Lattimore, C., Mura, P. and Yung, R. (2019) The time has come: A systematic literature review of mixed methods research in tourism. *Current Issues in Tourism* 22 (13), 1531–1550.

Kozak, M. and Rimmington, M. (1999) Measuring tourist destination competitiveness: Conceptual considerations and empirical findings. *International Journal of Hospitality Management* 18 (3), 273–283.

Krejcie, R.V. and Morgan, D.W. (1970) Determining sample size for research activities. *Educational and Psychological Measurement* 30, 607–610.

Lamont, M. and Jenkins, J. (2013) Segmentation of cycling event participants: A two-step cluster method utilizing recreation specialization. *Event Management* 17 (4), 391–407.

Le, D., Scott, N., Lohmann, G. (2019) Applying experiential marketing in selling tourism dreams. *Journal of Travel and Tourism Marketing* 36 (2), 220–235.

Leung, A., Yen, B.T.H. and Lohmann, G. (2017) Why passengers' geo-demographic characteristics matter to airport marketing. *Journal of Travel and Tourism Marketing* 34 (6), 833–850.

Li, H., Pearce, P.L. and Zhou, L. (2015) Documenting Chinese tourists' motivation patterns. Paper presented at the Rising Tides and Sea Changes: Adaptation and Innovation in Tourism and Hospitality: Proceedings of the 25th Annual CAUTHE Conference, Gold Coast, Queensland.

Lohmann, G. and Pearce, D.G. (2012) Tourism and transport relationships: The suppliers' perspective in gateway destinations in New Zealand. *Asia Pacific Journal of Tourism Research* 17 (1), 14–29.

Mazanec, J.A. (1984) How to detect travel market segments: A clustering approach. *Journal of Travel Research* 23 (1), 17–21.

Morgan, D. (2014) Pragmatism as a paradigm for social research. *Qualitative Inquiry* 20 (5), 1–9.

Munar, A.M. and Jamal, T. (2016) What are paradigms for? In A.M. Munar and T. Jamal (eds) *Tourism Research Paradigms: Critical and Emergent Knowledges* (pp. 1–16). Bingley: Emerald.

Murdy, S., Alexander, M. and Bryce, D. (2018) What pulls ancestral tourists 'home'? An analysis of ancestral tourist motivations. *Tourism Management* 64, 13–19.

Neuman, W.L. (2011) *Social Research Methods*. Boston: Pearson.

Pabel, A. and Coghlan, A. (2011) Dive market segments and destination competitiveness: A case study of the Great Barrier Reef ecosystem health. *Tourism in Marine Environment* 7 (2), 55–66.

Pansiri, J. (2006) Doing tourism research using the pragmatism paradigm: An empirical example. *Tourism and Hospitality Planning and Development* 3 (3), 223–240.

Pearce, P.L. (2005) *Tourist Behaviour: Themes and Conceptual Schemes*. Clevedon: Channel View Publications.

Pearce, P.L. and Lee, U.-I. (2005) Developing the travel career approach to tourist motivation. *Journal of Travel Research* 43 (3), 226–237.

Penrose, E. (1959) *The Theory of the Growth of the Firm*. New York: Wiley.

Poon, A. (1993) *Tourism, Technology and Competitive Strategies*. Wallingford: C.A.B. International.

Porter, M.E. (1990) *The Competitive Advantage of Nations*. New York: Free Press.

Powell, T.C. (2001) Competitive advantage: Logical and philosophical considerations. *Strategic Management Journal* 22 (9), 875–888.

Queiroz Neto, A., Lohmann, G. and Scott, N. (2016) Destination competitiveness: Theoretical gap, practical consequences. Paper presented at the CAUTHE 2016: The Changing Landscape of Tourism and Hospitality: The Impact of Emerging Markets and Emerging Destinations, Sydney.

Queiroz Neto, A., Lohmann, G., Scott, N. and Dimmock, K. (2017) Rethinking competitiveness: Important attributes for a successful scuba diving destination. *Tourism Recreation Research* 42 (3), 356–366.

Queiroz Neto, A., Scott, N., Lohmann, G. and Dimmock, K. (2018) Attributes, consequences and desired end-states of a successful scuba diving destination. CAUTHE

2018: Get Smart: Paradoxes and Possibilities in Tourism, Hospitality and Events Education and Research, pp. 32–45.
Spasojevic, B., Lohmann, G. and Scott, N. (2018) Air transport and tourism: A systematic literature review (2000–2014). *Current Issues in Tourism* 21 (9), 975–997.
Stockman, C. (2015) Achieving a doctorate through mixed methods research. *Electronic Journal of Business Research Methods* 13 (2), 74.
Sung, H.H. (2004) Classification of adventure travelers: Behavior, decision making, and target markets. *Journal of Travel Research* 42 (4), 343–356.
Tashakkori, A. and Teddlie, C. (1998) *Mixed Methodology: Combining Qualitative and Quantitative Approaches.* Thousand Oaks, CA: Sage.
Tashakkori, A. and Teddlie, C. (2003) *Handboook of Mixed Methods in Social and Behavioral Research.* Thousands Oaks, CA: Sage.
Vargo, S.L. and Lusch, R.F. (2004) Evolving to a new dominant logic for marketing. *Journal of Marketing* 68 (1), 1–17.
Wernerfelt, B. (1995) The resource-based view of the firm: Ten years after. *Strategic Management Journal* 16 (3), 171.
Wilde, S.J. (2010) A holistic investigation into principal attributes contributing to the competitiveness of tourism destinations at varying stages of development. PhD thesis. Southern Cross University, Australia.
Wilde, S.J., Cox, C., Kelly, S.J. and Harrison, J.L. (2017) Consumer insights and the importance of competitiveness factors for mature and developing destinations. *International Journal of Hospitality and Tourism Administration* 18 (2), 111–132.
Woodruff, R.B. (1997) Customer value: The next source for competitive advantage. *Journal of the Academy of Marketing Science* 25 (2), 139–153.
Woodruff, R.B. and Gardial, S.F. (1996) *Know your Customer: New Approaches to Understanding Customer Value and Satisfaction.* Cambridge, MA: Blackwell Business.

12 Pragmatism in the Context of Urban Design and Tourism: A Multi-disciplinary Study

Allison Anderson

Introduction

To create knowledge from observing things requires an acknowledgement of how you look at them. Traditionally in academic research, this has required the situation of a theoretical framework within disciplinary boundaries. This is a straightforward process when your research sits tidily within a single discipline. However, in studies that sit within interdisciplinary fields, or even multiple interdisciplinary fields, which draw from a range of disciplinary underpinnings, understanding and adopting a paradigm to work within can be challenging. You need to ask yourself: At what point does interdisciplinarity move to post-discipline? How do you acknowledge the complexity of context and disciplinarity while still shaping up a robust and manageable study for your PhD?

These were questions at the foundation of my research, which drew on two interdisciplinary fields of study: urban design and tourism (Goldstein & Carmin, 2006; Jafari & Brent Ritchie, 1981). A strength of both of these fields is the way they draw together different ways of looking at problems and phenomena from different disciplinary perspectives, but researching within and across them also presents methodological challenges. Early on in my PhD, I identified pragmatism as a paradigm that could transcend disciplines, while still respecting disciplinarity through addressing the 'wicked problems' inherent in tourism and urban design. Pragmatism enables the researcher to embrace a plurality of epistemological and so, methodological approaches that affords the flexibility that is required in addressing complex research questions (Johnson & Onwuegbuzie, 2004), liberating the researcher from the constraints of defined ontologies and epistemologies to focus on their methodology (Morgan, 2007).

In this chapter, I review the pragmatic approach to research and how it relates to traditional disciplinary structures. I then explain how pragmatism anchored my PhD study and conclude with a reflection on my experiences working with it.

The Pragmatic Approach

Pragmatic approaches to research have been around since the late 1800s; however, they have been gaining ground as an alternative to the traditional disciplinary approaches since the 1970s (Legg & Hookway, 2019). The Cambridge Dictionary (2016) describes pragmatism as 'the quality of dealing with a problem in a sensible way that suits the conditions that really exist, rather than following fixed theories, ideas or rules.' This general definition applies cleanly to the academic literature which argues that at the core of pragmatism is Peirce's (1999) Pragmatic Maxim, 'a rule for clarifying the meaning of hypotheses by tracing their "practical consequences" – their implications for experience in specific situations' (Legg & Hookway, 2019).

Building on these ideas of a problem focus within a specific context, Morgan (2007: 68) proposes a pragmatic approach that is centred on methodology, challenging the 'top-down privileging of ontological assumptions'. He argues that while acknowledging ontological and epistemological frameworks is important, it is 'simply too narrow an approach' (Morgan, 2007: 68) to be fully guided by them. In his later work, Morgan (2014: 1) presents pragmatism as a philosophy 'for understanding the nature of social research'. He contends that pragmatism serves to address inquiries about human actions and experiences, ranging from general inquiries to research. Morgan (2014: 7) argues that increasingly pragmatism responds to inquiries that have a social justice and/or political focus and that this presents 'pragmatism as a coherent philosophy that goes well beyond "what works" . . . and points to the importance of joining beliefs and actions in a process of inquiry'. Building on the notion of pragmatism as a paradigm, Legg and Hookway (2019) cite the importance of ontology in pragmatism, arguing that a key application of Peirce's (1999) maxim is clarifying the concept of truth. In other words, identifying how you view reality is a foundation of the pragmatic approach. This insight was important to my research as it was situated in the tropics, a region that has long been the subject of critical studies and analysis (Echtner & Prasad, 2003). As such, within the pragmatic paradigm I held a critical realist ontological viewpoint.

I was initially drawn to pragmatism in my research because it offered a path to focus on a practical problem that I knew existed, without getting too hung up on the 'fixed theories and rules'. Through previous work and experience, I knew that there was a disconnect between what people wanted out of the urban landscape in Cairns and what was being created.

But as I started my research, I quickly realised that although the problem looked straightforward, the foundations of the problem were quite the opposite. As I researched the two base fields of study – urban design and tourism, it became evident that these were full of 'wicked' problems that transcended disciplines and did not have a clear solution (McKercher, 1999; Rittel & Webber, 1973), and that there was no common disciplinary foundation for me to draw on in the traditional sense to guide an approach. Pragmatism gave me an anchor, or a 'mooring' through which to stay focused on the problem at the heart of my research without getting mired in disciplinary boundaries or the complexity of the wicked problems surrounding it.

Multidisciplinary Study and Post-discipline

As advances are made in knowledge, methodologies and technology, the development and re-shaping of disciplines is imperative. Increasingly, researchers are being faced with problems that transcend traditional disciplinary boundaries. Often, the differentiating factor of many projects emerging today is an ability to look beyond the traditional confines of the researcher's original discipline. These are usually described as 'multidisciplinary' or 'interdisciplinary' studies. Leiper (1981, 1990) defines 'multidisciplinarity' as combining multiple disciplines, while working within the confines of the disciplinary practices and codes. Identifying this concept as problematic and complex, Leiper suggests that 'interdisciplinarity' is a more usable concept, where studies sit between the disciplines, using what is required from the disciplines involved as needed. There exists a range of shades on the disciplinary spectrum, and many authors have attempted to produce definitive definitions and terms. Stember (1991: 5) produces a 'typology for enterprises within and across disciplines' gradually increasing in scope from intradisciplinary (within disciplines) to cross-disciplinary (looking at one discipline from the perspective of another), multidisciplinary (involving several disciplines each providing a perspective on a problem), interdisciplinary (integrating the contributions of several disciplines) and transdisciplinary (uniting intellectual frameworks beyond the disciplinary perspectives). More recently, Krishnan (2009) argues that interdisciplinarity occurs simply when disciplinary boundaries are crossed, and in fact encompasses all of these definitions (with the exception of intradisciplinarity). He goes on to suggest that the key lies not in defining the ever-expanding list of terms, but rather being able to identify when disciplinary boundaries are crossed – in essence, understanding the core of the issue rather than the edges.

Although the list of authors engaging in the debate on the levels of disciplinarity is long, there is general agreement on the use of interdisciplinarity as the concept of crossing disciplinary boundaries. The level of integration within this concept is still being debated; however, a number

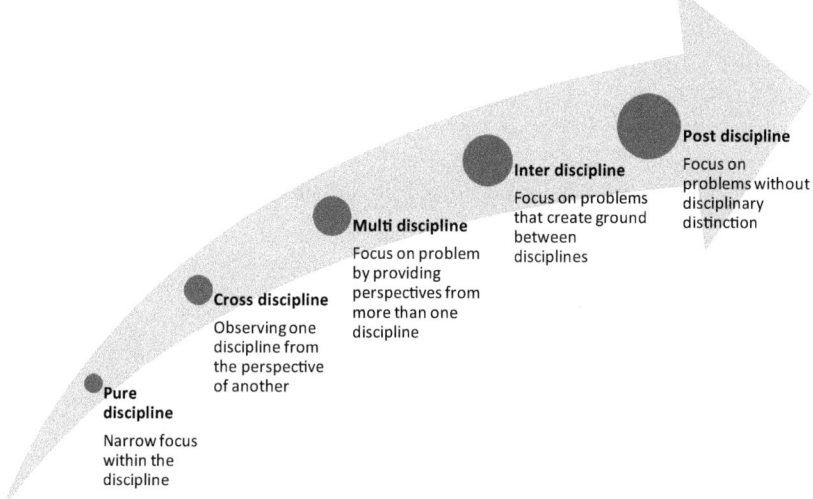

Figure 12.1 The changing focus of disciplinary research
Source: Adapted from Echtner & Jamal, 1997; Jafari & Ritchie, 1981; Sholle, 1995; Stember, 1991.

of authors in recent years have considered the movement beyond the disciplines to a 'post-disciplinary' approach away from the traditional ways of organising knowledge into disciplines and towards a more problem-focused approach that emphasises epistemological plurality. Figure 12.1 shows a simplified continuum of the broadening focus of disciplinary research, where further down the line from 'interdisciplinarity' would be 'post-disciplinarity'. Coles *et al.* (2006: 293) suggest that tourism studies in particular would 'benefit greatly from a post-disciplinary outlook… which is more problem-focused, based on more flexible modes of knowledge production, plurality, synthesis and synergy'. Sholle (1995: 141) presents much the same argument for media studies, suggesting that the 'walls of disciplinarity be replaced with bridges', Barnes (2006) for economic geography, and Painter (2003) for political geography. These arguments stress that they are not wishing to remove the organisation of knowledge altogether in an 'anti-discipline' approach, but rather promote an acceptance of the wide range of research that falls beyond the limits of the traditional disciplines through a 'post-discipline' approach.

As interdisciplinary fields are developing more rapidly, the academy's weighty disciplinary structure is struggling to keep up. For those researchers facing wicked problems in interdisciplinary research, the post-discipline concept can be appealing because of the aspects it promotes – problem solving, pragmatism, plurality and synergy. Krishnan (2009: 51) takes a longer view however, and considers what would be lost in the post-discipline approach, arguing that 'it is unlikely that a single post-disciplinary science could be possible or successful or even desirable'. In particular, Krishnan draws our attention to the value of situating claims to truth and worldviews within disciplines to enable them to make sense. It is here that the views of Krishnan and those of the post-disciplinarians who claim not to be anti-discipline seem to converge: in a place where the value, traditions and theories of a discipline are acknowledged within an interdisciplinary study, but which is not constrained by the imperialism or silo mentality of the disciplines.

The tenets of pragmatism in post-disciplinary study

An underpinning concept of the post-disciplinary approach is a focus on addressing research objectives rather than established disciplinary methodologies. Morgan (2007) discusses this, arguing for a transcendence of the metaphysical paradigm concept of research to propose a pragmatic approach that enables the researcher to move between qualitative and quantitative approaches (see Table 12.1). A tenet of this approach is acknowledging disciplinary and/or epistemological boundaries, but with a focus on achieving the research objectives. Table 12.1 identifies that in the qualitative approach, the connection of theory and data is characterised by *induction*, where theories are derived and built from themes emerging from the data, which is a subjective process. Conversely, the quantitative approach connects theory and data with *deduction*, where already formed theories are tested by the data in what some describe as an objective process, although it is acknowledged that every research process is laden with the values of the researcher and their processes. The pragmatic approach is described by Morgan (2007) as a 'middle ground' where both induction and deduction are used in an iterative process, so theories can emerge from the data and be tested. This iterative process is referred to in Table 12.1 as *abduction*.

I was encouraged to learn that abduction has previously been applied in design thinking contexts, identified by Dorst (2011) as a valuable way to focus on the 'core' of design issues, where the 'what' and 'how' are unknown but the desired result of 'value' is known. This is closely related to Carmona

Table 12.1 Pragmatic alternative to the key issues in social science research methodology

	Qualitative approach	Quantitative approach	Pragmatic approach
Connection of theory and data	Induction	Deduction	Abduction
Relationship to research process	Subjectivity	Objectivity	Intersubjectivity
Inference from data	Context	Generality	Transferability

Source: Morgan, 2007: 71

et al.'s (2003) call to focus on 'the heart' of problems rather than prescribe boundaries or edges. Equally, it is acknowledged that studies of the built environment are most effective when both qualitative and quantitative analysis are used together (Amaratunga *et al.*, 2002). These considerations of pragmatism resonated with what I was seeking to capture in my research.

Critically considering the process of urban design within context and across disciplinary and structural boundaries could be considered more useful than a phenomenological approach, which seeks only to describe a particular phenomenon. In the urban studies literature, Banai (1995: 570) argues that critical realism 'seeks to reveal causation, not just the manifestation of urban and regional development phenomena'. These considerations are echoed by Dovey (1999: 44), who argues that focusing on phenomenology and experience 'runs the risk that the ideological framings of place can remain buried and hence powerful'. Responding to Dovey's (1999) concerns, my PhD used a pragmatic approach with a critical framing to 'unpick the philosophies embedded in urban design' (Maitland & Smith, 2009: 186). Critical consideration of the urban design process and tourist experiences span across a number of disciplinary foundations. Although debate still exists if their status can be considered as a discipline or not, it is widely accepted that urban design and tourism are both interdisciplinary fields of study, with equally complex underlying disciplinary approaches, thus making them suitable to consider within a post-disciplinary framework.

My Research – A Critical Pragmatic Case Study

The aim of my research was to critically examine the process of urban design in a tropical city in terms of its alignment with tourist preferences. Within this, I had four objectives:

(1) to analyse the place, polity and power contexts of urban design in a tropical city;
(2) to critically analyse the urban design process in a tropical city from the perspective of those shaping it;
(3) to identify the connections and disconnections between how urban designers shape a tropical city and what tourists value in the tropical city experience; and
(4) to propose a method of evaluating the urban design process in terms of its alignment with tourist preferences.

My research explored the process of urban design in a descriptive sense, but in considering its effectiveness in a specific context also considered the concept of tropical urban design in a normative sense. A normative approach extends descriptive studies by going beyond describing what *is*, to considering what *should be*, as articulated within the values of those who are involved or affected by the outcomes. This approach is supported by Loukaitou-Sideris (2012) who argues that in the struggle for good urban environments, urban designers need to test urban design outcomes. Although often neglected, testing outcomes is intrinsic to the concept of understanding and evaluating any process of design, as it is inevitably a process concerned with improving value, an inherently subjective concept. To illustrate this, Carmona (2003: 74) defines the normative approach in urban design as 'the process of making better places for people than would otherwise be produced'. The concept of what is 'better' or 'worse' in place experience is open to interpretation, however it is clear that there are places people generally like to be more than others.

My research critically examined the urban design process in the tropical Australian city of Cairns. I achieved this through first investigating the place, power and polity elements of the Cairns context, then through exploring the process from the perspective of those within the process and finally through analysing the effectiveness of the process in in terms of its alignment with tourist preferences.

The underlying ontological assumption in my research was that there is a reality that is possible to know, but that reality is socially constructed and needs to be critically considered. Guba and Lincoln (2005) identify that this knowable but subjective reality is shaped by a range of social, economic, political, cultural and other values. Previous research shows that the concept of tropicality in particular, has mediated the way in which people see a place as peripheral, 'other', 'uncivilised' or 'uncultured' (Echtner & Prasad, 2003), and that these enduring ways of thinking influence perceptions of urban designers and tourists (Kravanja, 2012; Luckman et al., 2009). My research critically considered the process of urban design as a reality within a specific context from the perspective of those who are involved in it, and from the perspective of those who experience the results of it. The critical approach sought to unearth hidden elements that are fundamental to the reality yet not necessarily known or acknowledged by those who are a part of the process. Table 12.2 shows a summary of the research approach I used.

Table 12.2 Research approach

Paradigm	Pragmatism		
Ontology	Critical realist		
Epistemology	Subjectivist		
Research approach	Deductive and inductive reasoning	Abductive reasoning	Inductive reasoning
Research methodology	Case study with abductive analysis		
Data collection methods and analyses	Context analysis Interviews Policy analysis Tripadvisor analysis	Semi-structured interviews	Questionnaire Statistical analysis Website analysis

The critical approach also requires an acknowledgement of personal perspective in the research (Ateljevic *et al.*, 2005). As a resident of Cairns who had worked in both the tourism and planning industries, my personal approach was informed by an innate knowledge of 'how things are done' in the city from previous experience. Although best efforts were made to take the stance of an independent observer to elicit an objective critique of the urban design process, I acknowledged that the research design, data collection and analysis bore some influence of my personal beliefs and perceptions. This was particularly evident in the identification and interviewing of urban designers for the research.

The pragmatic, critical case study nature of my research informed a methodology that used mixed methods analysed in deductive, inductive and abductive processes. Using Yin's (2014) typology of case study design, my research was a single case study with two embedded units of analysis. The case was defined as the urban design process in Cairns, with the two key units of analysis being the urban designers and the tourists. Within this design, Yin also notes the critical role of context. According to Yin's (2014: 52) definitions, the case study's objective 'is to capture the circumstances and conditions of an everyday situation... because of the lessons it might provide about the social processes related to some theoretical interest'.

My data collection used a mixed methods approach with a fully mixed sequential equal status design as described by Leech and Onwuegbuzie (2009), annotated as QUAL- > QUAL- > QUANT. Although in line with the pragmatic approach, the research was less sequential and more iterative, or abductive. *Abduction* was a central aspect of the pragmatic approach of my research. In line with this approach, the research was undertaken in five stages at times concurrently and other times sequentially consistent with the research objectives.

- Stage 1: Context analysis. This included a desktop analysis of the historical and policy context in Cairns using current and past strategies, planning documents and online historical records; analysis of promotional material both online and as tourist collateral to assess the presence of the urban form in promoting the city as a tourism destination; analysis of current industry based visitor research; and ongoing collection and analysis of Tripadvisor.com data on the top five 'Things to do in Cairns' for the duration of the research period.
- Stage 2: Interviews with urban designers. 20 semi-structured interviews of 1–2 hours with urban designers in Cairns were conducted, which were coded and analysed through an inductive process to elicit historical, political, and other contextual information about Cairns, to identify the design priorities of urban designers in terms of tourism and the tropics.
- Stage 3: Tourist surveys. Drawing from the findings in Stage 1 and 2 of the research, a survey was designed. A total of 548 research participants

completed the structured self-completion surveys which was administered in popular tourist areas in Cairns for four months over the high and low seasons. Qualitative and quantitative responses were analysed using traditional, visual and mapping methods, both inductively and deductively.
- Stage 4: Analysis and writing. Results of the analyses in Stages 1–4 were drawn together to evaluate the connections and disconnections in the supply and demand of urban design in the tropical city from a tourism perspective.

Reflection on the Pragmatic Approach

In commencing my PhD, my perception was that there needed to be more research conducted on the needs and preferences of tourists in the tropical urban landscape to better inform the urban design process. While this remains true – since understanding the complexities of the needs of key users in the urban environment can never be a bad thing – what emerged from my research as one of the key findings was (the realisation of) only some disconnection between what urban designers think tourists want and what they actually want. A more evident disconnection emerged between the design norms of urban designers and what outcomes in the city they are able to achieve within the place, polity and power contexts of the city. In other words, it seems that the urban designers mostly understood what tourists want from the tropical city experience, but there are barriers to achieving these outcomes that are context-specific. These were unexpected findings, but were enabled by the pragmatic, abductive research design I employed.

I would recommend using pragmatism as a paradigm for approaching complex problems across multiple fields of interdisciplinary study. In reflecting on my experiences, I have four key points of advice that I would offer anyone considering using this approach in their research.

(1) Stay focused on the heart of the problem. Critically important in stepping in to a chaotic or complex problem is the ability to remain clear on what it is you are trying to achieve. Rather than trying to order it, or solve all of the problems, I took Carmona et al.'s (2003) advice to focus on 'the heart' of the problem, rather than prescribe its edge or boundary. Understanding and articulating what the heart of the problem is early on in the research process will make this much easier. While the rigour of academia requires definitions of terms and concepts, it is helpful to draw on the idea that 'knowing something when you see it' might actually be enough, and painstakingly defining where the line is between what things are and are not, may distract you from deeply understanding a complex problem. In my research, concepts such as authenticity in the tourism product, the components of place

in place attachment, and in fact many of the urban problems in general are all contested and have whole theses already devoted to their definition. At many points in my thesis, I referred to the philosophy of sticking to the heart of the problem to help remain focused.

(2) Embrace fluidity. The sibling to staying focused is embracing fluidity. A helpful visualisation for this might be a moored yacht, which can face many different directions, but still remain centred on the one buoy, which is tied to the bottom of the seabed by something very heavy and does not move. In pragmatic research, the focus is the mooring, and the fluidity is the many ways in which the boat can move while tied to the buoy. In my research, I knew that there was a disconnect between what people wanted out of the urban landscape in Cairns and what was being created. This was my mooring. As I mentioned, I thought this was because the urban designers did not understand the needs of the people using the landscape, but it was in fact for a different reason altogether. The iterative nature of the abductive approach I used, where I went between qualitative and quantitative research methods to evolve my thinking built fluidity, and helped me look at the problem in many different ways until I felt I had captured the heart of the problem.

(3) Pick your supervisors carefully. One of the workshops my university provided to PhD students identified that many of us are perfectionists, or at least high achievers. I saw evidence of this particularly in how I related to my supervisors, where I was seeking clarity and feedback, but instead generally received inconsistency and ambiguity. When my supervisors said – 'you need to look further into this' I would feel happy, and full of purpose. However, the very next meeting, when I proudly showed them evidence of my toils and they would proclaim 'why on earth are you wasting your time going down that path? You need to stay focused!' I would feel bereft, lacking motivation and solidly questioning why I ever thought I was cut out to do a PhD.

I realised over time that my supervision team had very different styles, paradigmatic perspectives and academic approaches. Until the very end of my research, this did not change, and with no clear leader in the team, I often found myself pulled between differing viewpoints and approaches. From my perspective, although I chose people I liked and respected for my supervision team, I was unprepared for the politics of academia and had not accounted for how important it was that these people respect each other and provide consistent advice. If you are considering using a pragmatic, problem-focused approach, I strongly recommend securing a lead supervisor who deeply understands how to chart that kind of research project.

(4) Context is critical, central even. One of the big lessons I learnt from place-based pragmatic research is the importance of understanding context. Authors such as Hall (2003), Carmona (2014) and Peck and Theodore (2010) have argued that these dimensions of context are

central to urban or tourism development. Indeed, one of Rittel and Webber's (1973) ten characteristics of wickedness is that each wicked problem is unique, suggesting that context is central when considering wicked problems in a critical way. I originally saw the context analysis in my thesis as perfunctory in the research process, as simply a stage-setting exercise to inform the reader of the specificities of Cairns as a place. However, it transpired that understanding the context ultimately led me to the underlying issues that built most of my findings.

Conclusion

This chapter has discussed the several complexities of using pragmatic research in multidisciplinary study. It introduced some ways to think about disciplinarity and considered how pragmatic research is well suited to projects that are seeking to improve messy, complex and 'wicked' real-world problems with ambiguous disciplinary underpinnings. It described how, while pragmatism is valuable in embracing diversity in approaches and methodologies, it is still valuable to identify an ontological viewpoint from the outset. It identified how a mix of inductive and deductive methods can be used iteratively in *abduction* to remain focused on the problem but also agile and fluid in how you consider it. I detailed how I designed my research within this structure, and then ruminated on the effectiveness of the approach.

For some researchers, the selection of a paradigm, ontology, epistemology and the research methods that support these, is simple, clean and straightforward. Their research deliberately focuses on one specific problem, and is often academically driven. While I envied those people their simplicity and clarity in approach, I recognise now how important my journey was in the research process. I sought to address a tricky and complex problem in fields of research that are, essentially, 'mongrels' with no clear single disciplinary underpinnings. I stepped into the chaos. The pragmatic research journey gave me the ability to deal with complex, wicked problems that have no clear solution, which is the true nature of many issues in the world today. These problems require creativity, discipline and focus, and are more likely to be 'improved' rather than 'solved'. Although my journey was stressful because of the ambiguity and need to regularly adapt research strategies and plans, I believe that there is particular value in this – not only for the world in benefiting from real-world problems being improved, but also for the researcher in their future career.

References

Amaratunga, D., Baldry, D., Sarshar, M. and Newton, R. (2002) Quantitative and qualitative research in the built environment: Application of "mixed" research approach. *Work Study* 51 (1), 17–31. doi:10.1108/00438020210415488

Ateljevic, I., Harris, C., Wilson, E. and Collins, F.L. (2005) Getting 'entangled': Reflexivity and the 'critical turn' in tourism studies. *Tourism Recreation Research* 30 (2), 9–21. doi:10.1080/02508281.2005.11081469

Banai, R. (1995) Critical realism and urban and regional studies. *Environment and Planning B. Planning and Design* 22 (5), 563–580.

Barnes, T.J. (2006) Situating economic geographical teaching. *Journal of Geography in Higher Education* 30 (3), 405–409. doi:10.1080/03098260600927211

Carmona, M. (2014) The place-shaping continuum: A theory of urban design process. *Journal of Urban Design* 19 (1), 2–36. doi:10.1080/13574809.2013.854695

Carmona, M., Heath, T., Tiesdell, S. and Oc, T. (2003) *Public Places-Urban Spaces: The Dimensions of Urban Design*. Oxford: Architectural Press.

Cambridge Dictionary (2016) Pragmatism. Cambridge University Press. Retrieved from: https://dictionary.cambridge.org/dictionary/english/pragmatism

Coles, T., Hall, C.M. and Duval, D.T. (2006) Tourism and post-disciplinary enquiry. *Current Issues in Tourism* 9 (4), 293–319. doi:10.2167/cit327.0

Dorst, K. (2011) The core of 'design thinking' and its application. *Design Studies* 32 (6), 521–532. doi:10.1016/j.destud.2011.07.006

Dovey, K. (1999) *Framing Places: Mediating Power in Built Form*. Abingdon: Routledge.

Echtner, C.M. and Jamal, T.B. (1997) The disciplinary dilemma of tourism studies. *Annals of Tourism Research* 24 (4), 868–883.

Echtner, C.M. and Prasad, P. (2003) The context of third world tourism marketing. *Annals of Tourism Research* 30 (3), 660–682. doi:http://dx.doi.org/10.1016/S0160-7383(03)00045-8

Goldstein, H.A. and Carmin, J. (2006) Compact, diffuse, or would-be discipline?: Assessing cohesion in planning scholarship, 1963–2002. *Journal of Planning Education and Research* 26 (1), 66–79. doi:10.1177/0739456x05282353

Guba, E.G. and Lincoln, Y.S. (2005) Paradigmatic controversies, contradictions, and emerging confluences. In N.K. Denzin and Y.S. Lincoln (eds) *The Sage Handbook of Qualitative Research* (3rd edn, pp. 191–215). London: SAGE Publications.

Hall, C.M. (2003) Politics and place: An analysis of power in tourism communities. In S. Singh, D.J. Timothy and R.K. Dowling (eds) *Tourism in Destination Communities*. Wallingford: CABI Publishing.

Jafari, J. and Brent Ritchie, J.R. (1981) Toward a framework for tourism education: Problems and prospects. *Annals of Tourism Research* 8 (1), 13–34. doi:10.1016/0160-7383(81)90065-7

Johnson, R.B. and Onwuegbuzie, A. (2004) Mixed methods research: A research paradigm whose time has come. *Educational Researcher* 33 (1), 14–26.

Kravanja, B. (2012) On social inequality in tourism development and tourist marketing of postcolonial Sri Lanka. *Studia ethnologica Croatica* 24, 107–129.

Krishnan, A. (2009) 'What are Academic Disciplines? Some Observations on the Disciplinarity vs. Interdisciplinarity Debate'. National Centre for Research Methods NCRM Working Paper Series 03/09. Southampton: University of Southampton.

Leech, N. and Onwuegbuzie, A. (2009) A typology of mixed methods research designs. *Quality and Quantity* 43 (2), 265–275. doi:10.1007/s11135-007-9105-3

Legg, C. and Hookway, C. (2019) Pragmatism. *The Stanford Encyclopedia of Philosophy*. Retrieved from https://plato.stanford.edu/archives/spr2019/entries/pragmatism/

Leiper, N. (1981) Towards a cohesive curriculum tourism: The case for a distinct discipline. *Annals of Tourism Research* 8 (1), 69–84. doi:10.1016/0160-7383(81)90068-2

Leiper, N. (1990) Tourism systems. Occasional Papers. Department of Management Systems, Massey University, Palmerston North, New Zealand, 2.

Loukaitou-Sideris, A. (2012) Addressing the challenges of urban landscapes: Normative goals for urban design. *Journal of Urban Design* 17 (4), 467–484. doi:10.1080/13574809.2012.706601

Luckman, S., Gibson, C. and Lea, T. (2009) Mosquitoes in the mix: How transferable is creative city thinking? *Singapore Journal of Tropical Geography* 30, 70–85. doi:10.1111/j.1467-9493.2008.00348.x

Maitland, R. and Smith, A. (2009) Tourism and the aesthetics of the built environment. In J. Tribe (ed.) *Philosophical Issues in Tourism* (pp. 171–190). Bristol: Channel View Publications.

McKercher, B. (1999) A chaos approach to tourism. *Tourism Management* 20 (4), 425–434. doi:http://dx.doi.org/10.1016/S0261-5177(99)00008-4

Morgan, D.L. (2007) Paradigms lost and pragmatism regained: Methodological implications of combining qualitative and quantitative methods. *Journal of Mixed Methods Research* 1 (1), 48–76. doi:10.1177/2345678906292462

Morgan, D. (2014) Pragmatism as a paradigm for social research. *Qualitative Inquiry* 20 (5), 1–9.

Painter, J. (2003) Towards a post-disciplinary political geography. *Political Geography* 22 (6), 637–639. doi:10.1016/s0962-6298(03)00070-2

Peck, J. and Theodore, N. (2010) Mobilizing policy: Models, methods, and mutations. *Geoforum* 41 (2), 169–174. doi:http://dx.doi.org/10.1016/j.geoforum.2010.01.002

Pierce, C.S. (1999) *The Essential Peirce* (Vol. 2). Bloomington: Indiana University Press.

Rittel, H.W. and Webber, M.M. (1973) Dilemmas in a general theory of planning. *Policy Sciences* 4 (2), 155–169.

Sholle, D. (1995) Resisting disciplines: Repositioning media studies in the university. *Communication Theory* 5 (2), 130–143. doi:10.1111/j.1468-2885.1995.tb00102.x

Stember, M. (1991) Presidential address: Advancing the social sciences through the interdisciplinary enterprise. *The Social Science Journal* 28 (1), 1–14.

Yin, R.K. (2014) *Case Study Research: Design and Methods* (5th edn). Los Angeles: SAGE.

# 13 In Search of an Intermediate Paradigmatic Ground: Critical Realism-Post-Positivism in Understanding Tourists' Motivation and Experiences in Asian Spas

Jenny H. Panchal

Introduction

Doing research was one of the things that I enjoyed the most, especially as an undergraduate tourism student in the Philippines (my home country). I took research classes, but I cannot remember any of my research professors discussing paradigms. Maybe if they did, a very small number of people would have retained them. Twenty years ago, in the Philippines, the concept of research was more of an educational, political or commercial requirement for marketing purposes. Like many other students and research organisations in the Philippines, I was more concerned about my research aims, data collection and analysis, and writing up of results. This mindset, which I call '*a-paradigmatic* (lack of paradigm) approach' to research, continued even after I learned about paradigms in Professor Douglas Pearce's research methods class in New Zealand and after completing a masters by thesis degree.

My *a-paradigmatic* mindset was shaped not by the actual absence of a research perspective, but the anxiety that paradigms and discussions about it caused me. Innumerable studies in psychology suggest that we often fear things that are unknown or unfamiliar to us. In this case, I knew and understood the paradigms; the unknown, however, was my own paradigm. As a new and young researcher at that time, it seemed that the intellectual debates and rejoinders on research paradigms were compelling me to choose a 'tribe' and work according to 'tribal rules'. Thinking about it caused stress, lack of appreciation of and a non-application of a paradigm in my research. I chose not to be part of any of the tribes.

Later, I learned that I cannot conduct research without a paradigm. Although I embarked on a PhD with an *a-paradigmatic* mindset, I had to come to terms with my view of the world, my perception of reality and how I value knowledge. Three months into my PhD, my supervisor, Professor Philip Pearce, invited me to an Honours class where he lectured on paradigms and research methods. As I listened to him, I was encouraged to identify my research identity and to be confident with it while creating and pursuing new knowledge. About two years later, after he had marked a high-quality PhD thesis with a very insightful section on paradigms, Professor Pearce reinforced the need to engage with my paradigm and to write a section on research paradigms and perspectives in my thesis.

My PhD thesis was entitled 'The Asia spa: A study of tourist motivations, flow and the benefits of spa experiences'. It aimed to highlight the relationship between tourism and positive psychology in the context of spa tourism in South East Asia by understanding tourist motivations, the experience of flow and the perceived benefits of spa-going while travelling. From an *a-paradigmatic* viewpoint, I struggled to ask myself: 'who am I as a researcher?' I knew I had to start somewhere, somehow. It was not until I was two years into my PhD journey that I began to understand who I am as a researcher and to establish the true direction of my PhD research.

As I went through literature in an attempt to identify my research paradigm, I found that my research objectives, methods and intended outcomes were not matching the paradigms I first thought shaped my research. Initially, I thought I was purely a post-positivist person (Guba & Lincoln, 1994, 2005) because of the data collection methods that I was using, but I also wanted to utilise a qualitative study with non-post-positivist characteristic. And then, I thought I was a pragmatist, because of my mixed methods approach and a dual research position, i.e. subjectivist and objectivist (Guba & Lincoln, 1994, 2005). How I viewed reality, however, was not in the way of a pragmatist. Finally, I decided that my worldview was in the middle ground, and that my paradigm was a hybrid one – a combination of at least two paradigms. And this, my friends, is what this chapter is about.

This chapter aims to underpin the importance of having a paradigm, to provide an understanding of the benefits of an 'intermediate paradigmatic ground', and to share accounts from the life of a PhD student who struggled to find her own research identity. This chapter consists of three parts: (a) an overview of the PhD, including key concepts used; (b) a brief review of paradigms in tourism and the use of an intermediate paradigmatic ground in my research; and (c) my post-PhD reflections about my research perspective.

The PhD: Overview and Context

The PhD project involved three studies, in which I tried to maintain consistency in terms of informants and geographical context. Study 1 was an on-site survey in India, Thailand and the Philippines. The questions were about 'flow' and travel career pattern (TCP) scales, previous travel and spa experiences and other profiling information. The respondents were spa-going tourists at the destination. Study 2 was an online survey that was similar to the on-site survey, except for modifications, which were mainly rating statements of the TCP statements based on two levels: general travel motives and spa-going motives. The online survey also did not include the flow scales. Lastly, Study 3 involved netnography, that is, an analysis of travel blogs that contained accounts of tourists' spa experiences in South East Asia, specifically in the countries where Study 1 was conducted.

This plan, however, was not the original one. I was supposed to explore motives and experiences of a narrow segment of tourists, i.e. those who were staying in spa or health resorts/retreats for wellness programs, and not a general population of spa-goers. A special access to spa/wellness tourists was a major requirement for the study, but because privacy is an important factor for hotel and resort guests, I had to acquire verbal and written permission and/or sign declarations regarding protocols. So, I embarked on my field trip, and my first stop was India. I was scheduled to fly from Townsville, Australia to Mumbai via Hong Kong on 25 November 2008, and to meet with a spa manager of an upscale resort in South Mumbai on the afternoon of 26 November to discuss data collection protocols. The flight from Townsville, however, was delayed by several hours, affecting all connecting flights and meetings set for the next day. Knowing that I would likely miss the meetings, I asked to reschedule them later that week. I arrived at midday in Mumbai on 26 November, and on the evening of that day, the horrific terrorist attacks in South Mumbai took place.

Because of the attacks, I was left with very few potential respondents as most of the establishments I previously contacted withdrew my access to resort guests as part of their safety and security measures for guests and employees. It was also assumed that tourists felt vulnerable to security or privacy breaches considering that travel and other profiling information were to be collected from them. Three weeks after the attacks and without success in accessing potential respondents, I shared my frustration with Professor Pearce. He was optimistic that we could still carry out the research; we just had to modify the study's direction. The thought of going back to the drawing board, changing research objectives and, ultimately, changing travel plans to other parts of India, to Thailand and the Philippines frustrated me further. Additionally, I was still traumatised by the thought that I *could have been* in a meeting in that resort when the attacks happened. The calm encouragement and wise counsel that

Professor Pearce imparted amidst my challenges, however, helped me maintain my composure, balance and a sense of direction.

Concepts used in the PhD

Although the course of my project changed, it took me one and half years after the change to find my research paradigm. I decided that mine is a hybrid – a mix of critical realism and post-positivism. In considering these paradigms, however, it is important to understand key concepts used in my PhD. These concepts are: (a) wellness; (b) tourist behaviour and motivation; and (c) the relationship between positive psychology and tourism. A fourth concept introduced here is the emic–etic distinction and their relationship to paradigms in the latter part of the chapter.

Concept 1 – Wellness

In this day and age, wellness is a common term that is widely used by the general public. In analysing wellness as a term and concept, common themes were drawn from the literature in various disciplines and specialisms. It was found that wellness is simultaneously a process and a state of being. It is multidimensional/holistic, subjective, relative and task-oriented (Panchal, 2012, 2013). As a multidimensional and holistic concept, wellness involves not only health and well-being but also happiness and satisfaction. Saracci (1997) argues that the wellness state relates more closely to happiness than to health. Further, Smith and Kelly (2006) suggest that concepts of happiness and health are different even though both encompass the term 'wellness'. While health can be self-assessed, it is usually measured objectively by traditional medicine. Well-being, happiness and satisfaction, on the other hand, are subjective and can change over a period of time (Ryan & Deci, 2001, 2008). Since wellness is broadly subjective and usually based on self-judgement, its manifestation is also dependent on the individual; this means the person has a sense of control over the degree of wellness that he/she wants to obtain (Cowen, 1991; Carruthers & Hood, 2004), and therefore it is task-oriented (Myers *et al.*, 2000).

Concept 2 – Tourist behaviour: Motivation & the Travel Career Pattern (TCP) theory

Tourist behaviour is an important aspect in studying tourism. I have found that Clawson and Knetsch's (1966) idea of travel behaviour is one of the simplest methods to explain this concept (in Pearce, 2005). They propose five phases: (1) pre-purchase or anticipation; (2) physical travel to the destination; (3) on-site experience; (4) return travel; and (5) extended recall and reflection. A further analysis of this approach suggests that this behaviour is a complex cycle rather than linear in nature, where phases may overlap each other at the same period of time, or a new cycle may emerge at any given phase.

A concept that is closely linked with tourist behaviour is 'motivation'. It is argued that motivation is a long-term (Pearce, 1982) and intrinsic process (Dann, 1977; Crompton, 1979; Hsu & Huang, 2008). Pearce (2005: 25) defines motivation as 'the total network of biological and cultural forces which give value and direction to travel choice, behaviour and experience'. Pearce and his colleagues' work on motivation evolved over the years. In this study, I used and slightly modified Pearce and Lee (2005) Travel Career Pattern (TCP) theory, which is based on Maslow's hierarchy of needs theory. The TCP approach suggests three layers of motives. The first layer consists of core motives that fundamentally influence travellers' decisions regardless of their travel experience; these are novelty, escape/relax and relationships. The second layer are moderately important motives which tend to focus on self-enhancement and host community contact needs. The third and least important layer is defined by specialist needs such as nostalgia, romance and adrenaline-based adventure. Broadly, the TCP suggests that tourists' motivational patterns change over their life-stages and/or with travel experience (Pearce, 2005; Panchal & Pearce, 2011).

Concept 3 – The Positive Psychology-Tourism linkage and 'Flow'

Following Seligman and Czikszenthmihalyi (2000), Pearce (2007: 3) defines positive psychology as a 'scientific study of positive emotions, characters strengths and positive institutions serving or concerned with human happiness and well-being'. In my contribution to Pearce and Filep's (2013) book, I offered two areas where positive psychology (along the lines of wellness) and tourism are linked: (a) the amount of 'flow' in tourist experiences; and (b) the perceived benefits of spa experiences. Flow is an optimal psychological state that is defined by Csikszentmihalyi (1975) as an experience that 'stems from people's perceptions of challenges and skills in given situations' (cited in Ellis *et al.*, 1994: 337). In my PhD, I used the Flow State Scale (or FSS-2) in measuring flow among tourists who had at least one spa experience while on holiday. This tool was chosen as a post-event assessment of flow, and to ascertain the respondent's particular peak experience (Panchal, 2012, 2013).

Concept 4 – Emic and etic perspectives

My PhD uses both emic and etic perspectives. In simple terms, emic is an insider's view of a phenomenon, while etic is an outsider's view. For my thesis, I explained that an emic researcher becomes an insider by taking the perspective of people who are engaging in the behaviour, and using the knowledge bases of the setting, the people and their explanations to describe the phenomenon. An etic researcher, on the other hand, becomes an outsider by providing the participants a worldview or perspective to which they respond (Pearce, 2005).

Paradigms in Tourism

Since Kuhn (1962) first used the term paradigm in his work *The Structure of Scientific Revolutions,* so much has been written about it. Moreover, controversies were raised resulting in the so-called *paradigm wars* and the proliferation of literature on paradigms. Paradigms have demarcated schools of thought (e.g. natural and social sciences), and many thinkers have identified the dominant paradigms that shape their respective discipline(s) and specialism(s). In the debates, scholars have offered insights to abate the rampant intellectual conflicts, if not to appease some of those involved in them. One approach was to consider paradigms as a continuum that reflect a researcher's discipline and philosophy (Burrell & Morgan, 1979; Hunt, 1992; Meredith *et al.*, 1989). Another way of dealing with conflicting beliefs and practices was the adoption of alternative paradigms and use of mixed research methods (Guba & Lincoln, 1994; Meredith *et al.*, 1989; Tashakkori & Teddlie, 1998, 2003; Teddlie & Tashakkori, 2003, 2012). I find Tashakkori and Teddlie's collective works to be the most notable example of the evolution of multiple-paradigm options that resulted from appraisals and debates (see Teddlie & Tashakkori, 2012).

At the time when I was reading and writing about paradigms for my thesis, I followed debates within the study of tourism. For example, Hollinshead (2004; cited in Gale & Botterill, 2005) identified four paradigms in tourism: positivism, constructivism, critical theory and postpositivism. I noticed that although mainstream tourism research often works within a positivist/post-positivist paradigm (Phillimore & Goodson, 2004; Walle, 1997), the rejection of positivism is also commonplace in this field of study (Franklin & Crang, 2001; Nash, 2001; Pearce, 2004; Rojek & Urry, 1997). Pearce (2004) argues that it is important to understand what is being rejected, and it should be recognised that positivism is a fundamental perspective in studying the natural sciences. He further warns, '[t]here is a danger, however, in the over-enthusiastic and simplistic rejection of positivism as the bad boy of scientific thinking. There can be a value in natural science style methods for investigating human behaviour and there is a role for generalisations about tourism and tourist behaviour even if they are not law-like' (2004: 61). Also, Jennings (2010: 58) observed that although more emphasis on positivist/post-positivist paradigms was given to research in the past, the recent years have seen a gradual shift in the employment of qualitative approaches to obtain 'deeper meanings people attribute to tourism and tourism experiences, events and phenomena'.

In more recent times, these debates are still prevalent. Nevertheless, I observed that the discourses on tourism research paradigms have advanced; they are much richer and more complex. The current literature still focuses on the nature of paradigms but also recognises the changing approaches in knowledge production in tourism studies. This evolution in debates suggests that tourism as a field of study is more mature and deserves much respect as a discipline. The more recent works of Tribe *et al.* (2015) and Munar and Jamal (2016a) offer an insightful analysis of the nature and consequences of paradigms in tourism studies. In Tribe *et al.*'s (2015) trialogue, the authors observe that Kuhn has not considered the notion of a paradigm as appropriate for the social sciences; they explained that paradigms offer detailed distinction between the natural and the social sciences. Tribe *et al.* advocate an analysis of paradigms as an implicit response to Kuhn's call for a similar and comparative study of paradigms in other fields of study. In the same work, Tribe *et al.* (2015: 30) broadly suggests that the study of tourism is based on a structured evolution that is typified by a 'less universal, more fuzzy, more speculative and rarely completely settled but rather fluid and on the move' nature. Munar and Jamal's (2016b) book, for example, is a collection of essays on tourism research paradigms focused on critical and emerging topics.

Intermediate paradigmatic ground: Post-positivism and critical realism

A paradigm is distinguished by the stance of its proponents in three fundamental ways: (1) ontology (the nature of reality/what I thought to be a reality); (2) epistemology (what is known and my research position in relation to the subjects being studied); and (3) methodology (the process of conducting research). Heron and Reason (1997, cited in Guba & Lincoln, 2005) suggest a fourth basis in distinguishing paradigms. They suggest axiology that involves the type of knowledge that is valued and how it is valued, and takes into account the role of values in the research processes. The major paradigms in contemporary social and behavioural sciences and their basic assumptions are presented in Table 13.1.

My PhD research perspective follows Walle (1997) and Jennings' (2010) view that one topic can be studied using different paradigms. In my PhD thesis, I wrote: 'This position, which can be termed as *"intermediate paradigmatic ground"* is derived from similarities, if not meshing of the two perspectives in terms of ontological stances, epistemologies, methodologies and even axiologies. I coined this term to reflect a middle ground or a mediating position between two different sets of perspectives or approaches (Gale & Botterill, 2005; Hollinshead, 2004; Jennings, 2010). The notion of mixed paradigms, however, is not novel. It is noteworthy that these two paradigms are classified as one in the works of Hollinshead (2004) and Gale and Botterill (2005). The latter authors, following Stockmann (1983), argue that critical realism is not simply post-positivist; it is anti-positivist. While such a notion is accepted in this work, post-positivism and critical realism are each presented as individual paradigms in the justifications concerning why and how this research was developed.

Table 13.1 The major research paradigms

	Paradigms					
	Positivism	Post-positivism	Critical realism	Critical theory	Constructivism/ interpretivism	Pragmatism
Ontology (the nature of reality/what is thought to be a 'reality')	Reality exists 'out there' and is driven by immutable natural laws and mechanisms.	Reality exists but can never be fully apprehended. It can only be incompletely understood.	Fallible truths are produced by social and historical circumstances.	Virtual reality shaped by social, political, cultural, economic, ethnic and gender values; crystallised over time.	Realities exist in the form of multiple mental constructions, socially and experientially based, local and specific, dependent for their form and content on the persons who hold them.	Pragmatic view of the world that what works is what is 'real' or true; hence the acceptance of external reality.
Epistemology (what is known and how one is positioned in relation to reality)	It is both possible and essential for the inquirer to adopt a distant, non-interactive posture. Values and other biasing and confounding factors are thereby automatically excluded from influencing the outcomes. (Dualist/Objectivist)	Objectivity remains a regulatory ideal, but it can only be approximated, with special emphasis placed on external guardians such as the critical tradition and the critical community. The possibility of researcher bias is acknowledged. (Modified objectivist)	Objectivity can be attained. The possibility of researcher bias is acknowledged. (Modified objectivist)	Values mediate inquiry. (Subjectivist)	Inquirer and inquired into are fused into a single (monistic) entity. Findings are literally the creation of the process of interaction between the two. (Subjectivist)	Experience emerges as a continual interaction between people and their environment; accordingly, this process constitutes both the subjects and objects of inquiry. (Both subjectivist and objectivist)

Table 13.1 Continued.

	Paradigms					
	Positivism	Post-positivism	Critical realism	Critical theory	Constructivism/ interpretivism	Pragmatism
Methodology (the process of acquiring knowledge, i.e. research)	Questions and/or hypotheses are stated in advance in propositional form and subjected to empirical tests (falsification) under carefully controlled conditions.	Emphasise critical multiplism. Redress imbalances by doing inquiry in more natural settings, using more qualitative methods, depending more on grounded theory and reintroducing discovery into the inquiry process.	Emphasise multiplism. Primarily quantitative but may use qualitative methods.	Eliminate false consciousness and energise and facilitate transformation.	Individual constructions are elicited and refined hermeneutically, and compared and contrasted dialectically, with the aim of generating one (or a few) constructions on which there is substantial consensus.	Mixed methods.
Axiology (what and how knowledge is valued)	Knowledge is propositional and of intrinsic value.	(the same as positivism)	Knowledge is propositional, of intrinsic value and a potential means to social emancipation.	Knowledge is propositional, transactional and a way to achieve social emancipation.	(the same as critical theory)	(the same as critical realism)

Source: (From PhD thesis (p. 47) and based on Greenwood & Levin, 2005; Guba, 1990; Guba & Lincoln, 2005; Jennings, 2010; Tashakkori & Teddie, 2003; Teddlie & Tashakkori, 2003).

The following sections consider how the studies are shaped by ontological, epistemological, methodological and axiological considerations' (Panchal, 2012: 48).

Mixed paradigms in studying spa tourist motivations and experiences

It is important to be reminded that my research was about tourist motivation for, experiences of and reflections about spa tourism. The work was pursued within post-positivist and critical realist paradigms but also attempted to consider carefully the emic and user perspectives in the design and conduct of the studies. This section will provide an analysis and justification for an intermediate paradigmatic ground from ontological, epistemological, methodological and axiological views. The linkage between emic–etic perspectives and these paradigms will also be presented. An overview of these two paradigms is shown in Table 13.2.

Ontology

The post-positivist recognises that there is a 'real' reality in the world but it is not fully understood. Such reality driven by laws needs to be checked, evaluated and negotiated. The critical realist views reality as being shaped by different factors and as being crystallised over time (Guba & Lincoln, 2005). These two stances mesh in the context of the flourishing health and wellness phenomena. While I did not seek to identify laws that create the perceived reality for this project, I recognised the value of historical foundations of concepts in understanding the phenomenon. It was clearly acknowledged that the evolution of wellness based on historical accounts was an important source of its complexity and richness. This complexity is due to the negotiability of the term and concept. As discussed in the previous section, wellness is a multidimensional phenomenon that can be defined, interpreted and projected in different disciplines, norms and even scenarios. Despite the complex nature of the term, the wellness industry exists and continues to grow globally.

The historical development of a concept contributes much to its clarity and most importantly acceptability. The evolution and development of tourist behaviour as a broad concept, for example, is important in how it is understood today. I have observed that many earlier works that contribute to the understanding of tourist behaviour are inclined towards post-positivism (cf. Ryan, 2000). The academic negotiations, debates and rejoinders in Harrill and Potts' (2002) three phases of tourist motivation models development, for example, are an indication of post-positivist's need for evaluation and negotiation about the perceived reality. As Goodson and Phillimore (2004) observed, positivism and post-positivism paradigms provide the context within which many tourism researchers operate.

Table 13.2 The current research paradigm (Modified from my PhD thesis (p. 49) and based on Guba, 1990; Guba & Lincoln, 2005; Jennings, 2010)

	Paradigms		Intermediate paradigmatic ground
	Post-positivism	Critical realism	
Ontology	-There is a 'real' reality out there but it can never be fully understood (that is external to the researcher) -Reality is driven by natural laws that can only be incompletely understood -Reality needs to be checked, evaluated and negotiated about ('Does it really exist?')	-Reality is shaped by social and historical circumstances -Reality is crystallised over time -Treating social structures as 'reality' as a result of past practices as 'reality' can be re-examined	There is a 'real' reality outside the observer that has been crystallised over time. My ability to understand it was imperfect and the need for examination to get the best understanding of it was recognised. Such reality needed to be evaluated.
Epistemology	-Objectivity remains a regulatory ideal, but it can only be approximated (absolute researcher objectivity is *unattainable*). -Observations are theory-laden and influenced by theory, but the researcher can construct theory inductively. -Special emphasis is placed on external guardians such as the critical tradition and the critical community.	-The possibility of research bias is acknowledged -Separating the researcher from what is being studied is a challenge -The credibility of our position in relation to what is known can be checked by interested communities	Absolute researcher objectivity was unattainable, and observations were theory-laden. The 'emic' approach embedded me as the researcher in the phenomenon being studied, and I was willing to have the credibility/trustworthiness of my epistemological stance be checked and evaluated by the critical community (i.e. the people involved in the study or other researchers).
Methodology	-Emphasise critical multiplism: Chiefly quantitative with some tendency to see qualitative methods as a precursor to quantitative methods -Researches in more natural settings using more qualitative methods -Depends more on grounded theory and reintroducing discovery into the inquiry process.	-Eliminate false consciousness and energise and facilitate transformation -Encourage reflection -Encourage conversation and dialogue -Question people's current experiences	It was acknowledged that a single perfect scientific method does not exist. Qualitative and quantitative methods can be compromised. In the study, quantitative approaches were used primarily but qualitative methods eliciting participants' ways of reflecting and knowing (i.e. blogs/texts) were also used. Research designs aimed at boosting emergent or new discovery.
Axiology	-Is propositional and of intrinsic value	-Is propositional and of intrinsic value -Potential means of social emancipation	In both paradigms, knowledge proposes facts/realities rather than beliefs, which makes it value-free and neutral. Although knowledge may be a potential driver of social change, it was not an ultimate goal of the project.

Epistemology

The key questions in epistemology are 'what is known?' and 'how one is positioned in relation to reality?' Guba (1990) suggests that post-positivists are modified objectivists whose objectivity remains a regulatory ideal, but it can only be approximated (cited in Guba & Lincoln, 2005). He further adds that critical realists are modified subjectivists where the possibility of research bias is acknowledged. Subjectivism is an epistemological stance where the knower and the known become inseparable (Guba & Lincoln, 2005).

The meshing of these two epistemological positions was manifested by the inherent approximations of the observations, and the inability to reach total objectivity yet openness of the researcher's position to be evaluated in terms of credibility (partly subjective). The application of post-positivist epistemology is evident in the use of tourism-related theories (e.g. motivation) and positive psychology models (e.g. flow, well-being) to guide my study of tourist spa-goers behaviour in Asia. However, the critical realist part of me acknowledged potential bias because I am: (a) a Southeast Asian woman who understands the differences in wellness practices in Asia; (b) a regular spa-goer at home and while travelling, which collectively mean that I have a wealth of experience; and (c) knowledgeable of different standards in spa therapies and spa management in Asia. Hence, the difficulty that I faced in separating myself from my observations also typified me as a critical realist.

Methodology

My methodological perspective followed the works of Jenning's (2010) and Rossman and Wilson's (1985). Following Ambercrombie *et al.* (1988), Jennings (2010) suggests that an empirical statement or theory is one which can be tested by evidence drawn from experience. Both post-positivist and critical realist paradigms emphasise methodological pluralism where both quantitative and qualitative methods are used. The motivations and flow experiences in an Asian spa context were measured quantitatively (using questionnaires) and statistical calculations were employed in the studies. However, the on-site survey (Study 1) and online questionnaires (Study 2) were limited to certain aspects of tourists' spa experience. Scholars suggest that experiences cannot be accessed directly by quantitative research (Caru & Cova, 2008; cited in Panchal, 2012, 2014). Netnography, i.e. blog analysis (Study 3) was used to obtain people's thoughts and feelings about their spa experience(s).

Also linked to this notion of pluralism is Rossman and Wilson's (1985) work that suggests a continuum that can be differentiated with the extent to which each regards how qualitative and quantitative styles can co-exist and can be used together (Onwuegbuzie & Leech, 2005). The three schools of thought are the purists, sensationalists and pragmatists (Figure 13.1).

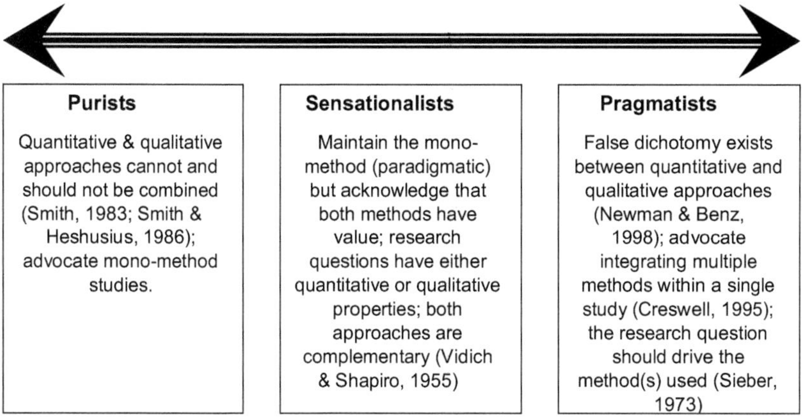

Purists	Sensationalists	Pragmatists
Quantitative & qualitative approaches cannot and should not be combined (Smith, 1983; Smith & Heshusius, 1986); advocate mono-method studies.	Maintain the mono-method (paradigmatic) but acknowledge that both methods have value; research questions have either quantitative or qualitative properties; both approaches are complementary (Vidich & Shapiro, 1955)	False dichotomy exists between quantitative and qualitative approaches (Newman & Benz, 1998); advocate integrating multiple methods within a single study (Creswell, 1995); the research question should drive the method(s) used (Sieber, 1973)

Figure 13.1 The differences between qualitative and quantitative research paradigms (Adapted from Onwuegbuzie & Leech, 2005: 376–377)

My PhD work used a *sensationalist perspective* where much value was given to both quantitative and qualitative approaches although a single method was employed in each study.

Axiology

Knowledge in both post-positivism and critical realism is proposed by facts and not by beliefs. In my thesis, I wrote: 'Such propositional knowing about the world is an end in itself (Guba & Lincoln, 2005). The critical realist's axiology, however, is extended to the possibility of shaping social emancipation. While such an objective is admirable, the topic for this research is narrow and does not extend significantly to social-change and policy-making imperatives' (Panchal, 2012: 51).

Emic and etic

As explained earlier, I am part of the spa-going populace in Asia, by which I considered myself an emic researcher. While collecting data for Study 1 specifically, I stayed in spa resorts, bought spa packages and spa treatments which were vastly available to tourists in India, Thailand and the Philippines. By taking part in activities that potential respondents were involved in, I had the opportunity to gain insight on different spa experiences. As an *insider*, I was challenged by segregating what I knew from my respondents' perspectives.

To address this critical realist epistemological position, I adopted an etic approach according to Fetterman (1989). He argues that a researcher may employ an emic perspective while collecting data, but must detach himself from the data at some stage to make sense of the empirical material. The outsider's view that I employed had two layers of detachment: (a) incorporating blog analyses which allowed me to explore the unknown among tourist bloggers; and (b) interpreting collected data as PhD student, and not as a fellow spa tourist.

There were episodes throughout my PhD where my emic/etic views were challenged. The most notable incidents occurred during the data collection in Thailand. The paper surveys were distributed in beaches, spas, resorts and departure areas in airports in Bangkok and Phuket, and as I and/or my local research assistant approached potential respondents, we were asked if we offered massage treatments as well. While some were generally curious, there were those who commented with sexual undertones. I tried my best to respond with tact and logic. With an emic mindset, however, it was challenging to ignore comments that I perceived were rude, condescending or insensitive. During the last two weeks of data collection in Thailand, I distributed the surveys by myself (i.e. without research assistants). At that time, I decided to see myself as a researcher rather than a tourist spa-goer. I noticed that with an etic perspective, I felt less offended in similar situations. I took note of the comments but did not report them in my thesis.

As I approached the end of the PhD, I was confident with an intermediate research perspective. I had mostly the ontological and epistemological functions of a post-positivist, and the methodological and axiological bases of a critical realist. I started with no clear paradigm and ended with a multi-paradigm mindset. I did not think much about paradigms in isolation for the rest of the candidature, but I was at peace knowing that I had good supervisors, who had a sound perspective of the world and wealth of opportunities to produce knowledge.

Post-PhD Musings on Paradigms

It's been over five years since I completed my PhD. When I look back, I smile; this may sound trite, but 'the entire PhD journey was not easy, but it was worth it'. On my graduation day, the guest of honour commented that my thesis sounded interesting and asked if I would have done it again. I said yes without batting an eyelash. Her follow-up question was, 'is there anything that you would have done differently?' I remember responding politely, but I could not recall what I told her. In writing this chapter, however, I had the opportunity to ponder the same question that I was asked on graduation day. And, in closing, I wish to impart my response to her question as if she were asking me again today, and I am hopeful that new researchers will find these insightful and encouraging.

An appreciation and application of at least one research paradigm is valuable in one's research work. I learned that knowing the paradigms *per se* is not enough; *knowing my paradigmatic identity in research* should have been one of my priorities. My a-paradigmatic approach to research when I began my PhD was a result of anxiety of maybe choosing a wrong paradigm. It took me years to realise that as a researcher, identifying one's self to a paradigm is better than not having a research perspective at all. Until recently, I blamed the 2008 Mumbai attacks for the change in my research direction. In retrospection, however, it is clear that my internal struggle was caused by a lack of recognition of my own research perspective.

I also realised that a mismatch in ontological stance, epistemological, methodological and axiological bases was not a sign of failure but an opportunity for intellectual growth. I could have recognised the 'intermediate paradigmatic ground' as an alternative to mono-paradigms much earlier in the PhD candidature. By having an intermediate perspective, I was able to logically link conclusions from my three studies because being in the 'middle ground' provided greater appreciation of the strengths and limitations of each paradigm as well as flexibility in designing research.

An intermediate paradigmatic ground also encouraged me to have both insider and outsider perspectives. Despite the challenges in data collection, analysis and interpretation, I appreciated the value of having emic and etic mindsets. Nevertheless, I could have *established a stronger etic approach earlier* in my research. In hindsight, I could have responded to questions and interpreted comments (e.g. sexual undertones) in Study 1 with a researcher mindset rather than a spa-goer mindset. From a methodological viewpoint, I could have used the help of research assistants throughout the data collection period in Thailand. Clearly, an emic perspective was more dominant, but an etic approach was useful in addressing epistemological issues and in diversifying research methods.

Indeed, Kuhn (1969) was right in describing paradigms as essential elements to scientific inquiry. He wrote, 'men whose research is based on shared paradigms are committed to the same rules and standards for scientific practice' (cited in Sarantakos, 1998: 32). I find that the analogy of choosing a tribe is still a fitting way of identifying one's paradigm, and one of the best things that resulted from the decades of debates is the option of an intermediate paradigmatic ground, which allows researchers to be part of more than one tribe.

References

Abercrombie, N., Hill, S. and Turner, B.S. (1988) *The Penguin Dictionary of Sociology* (2nd edn). Harmondsworth: Penguin Books.
Burrell, G. and Morgan, G. (1979) *Sociological Paradigms and Organisational Analysis*. Ashgate: Aldershot.
Carruthers, C.P. and Hood, C.D. (2004) The power of the positive: Leisure and well-being. *Therapeutic Recreation Journal* 38 (2), 225–245.
Clawson, M. and Knetsch, J.L. (1966) *Economics of Outdoor Recreation*. Baltimore: John Hopkins Press.
Cowen, E.L. (1991) In pursuit of wellness. *American Psychologist* 46, 404–408.
Creswell, J.W. (1995) *Research Design: Qualitative and Quantitative Approaches*. Thousand Oaks, CA: Sage.

Crompton, J. (1979) Motivations for pleasure travel. *Annals of Tourism Research* 6, 409–424.
Csikszentmihalyi, M. (1975) *Beyond Boredom and Anxiety*. San Francisco: Jossey Bass.
Dann, G. (1977) Anomie, ego-enhancement and tourism. *Annals of Tourism Research* 4, 184–194.
Ellis, G.D., Voelkl, J.E. and Morris, C. (1994) Measurement and analysis issues with explanation of variance in daily experience using the flow model. *Journal of Leisure Research* 26 (4), 336–356.
Fetterman, D.M. (1989) *Ethnography: Step by Step* (Vol. 17). Newbury Park: SAGE.
Franklin, A. and Crang, M. (2001) The trouble with tourism and travel theory. *Tourist Studies* 1 (1), 5–22.
Gale, T. and Botterill, D. (2005) A realist agenda for tourist studies, or why destination areas really rise and fall in popularity. *Tourist Studies* 5 (2), 151–174.
Goodson, L. and Phillimore, J. (2004) The inquiry paradigm in qualitative tourism research. In J. Phillimore and L. Goodson (eds) *Qualitative Research in Tourism: Ontologies, Epistemologies and Methodologies* (pp. 30–45). London: Routledge.
Greenwood, D.J. and Levin, M. (2005) Reform of the social sciences and of universities through action research. In N.K. Denzin and Y.S. Lincoln (eds) *The Sage Handbook of Qualitative Research* (3rd edn, pp. 43–64). London: SAGE Publications.
Guba, E.G. (1990) The alternative paradigm dialog. In E.G. Guba (ed.) *The Paradigm Dialog* (pp. 17–27). London: SAGE Publications.
Guba, E.G. and Lincoln, Y.S. (1994) Competing paradigms in qualitative research. In Denzin, N.K. and Y.S. Lincoln (eds) *Handbook of Qualitative Research*. Thousand Oaks, CA: Sage Publications.
Guba, E.G. and Lincoln, Y.S. (2005) Paradigmatic controversies, contradictions, and emerging confluences. In N.K. Denzin and Y.S. Lincoln (eds) *The Sage Handbook of Qualitative Research* (3rd edn, pp. 191–215). London: SAGE Publications.
Harrill, R. and Potts, T.D. (2002) Social psychological theories of tourist motivation: Exploration, debate and transition. *Tourism Analysis* 7, 105–114.
Hollinshead, K. (2004) A primer in ontological craft: The creative capure of people and places through qualitative research. In J. Phillimore and L. Goodson (eds) *Qualitative Research in Tourism: Ontologies, Epistemologies and Methodologies* (pp. 63–82). London: Routledge.
Hsu, C.H.C. and Huang, S.S. (2008) Travel motivation: A critical review of the concept's development. In A.G. Woodside and D. Martin (eds) *Tourism Management: Analysis, Behaviour and Strategy* (pp. 14–27). Oxford: CAB International.
Hunt, S.D. (1992) For reason and realism in marketing. *Journal of Marketing* 56 (2), 89–102.
Jennings, G. (2010) *Tourism Research*. Milton, Qld: John Wiley & Sons Australia.
Kuhn, T.S. (1962) *The Structure of Scientific Revolutions*. Chicago: University of Chicago Press.
Kuhn, T.S. (1969) *Postscript to the Second Edition of the Structure of Scientific Revolutions*. Chicago: University of Chicago Press.
Meredith, J.R., Raturi, A., Amoako-Gyampah, K. and Kaplan, B. (1989) Alternative research paradigms. *Journal of Operations Management* 8 (4), 297–326.
Munar, A.M. and Jamal, T. (2016a) What are paradigms for? In A.M. Munar and T. Jamal (eds) *Tourism Research Paradigms: Critical and Emergent Knowledge* (pp. 9–21). Bradford: Emerald Group Publishing.
Munar, A.M. and Jamal, T. (2016b) *Tourism Research Paradigms: Critical and Emergent Knowledge*. Bradford: Emerald Group Publishing Ltd.
Myers, J., Sweeney, T. and Witmer, J. (2000) The wheel of wellness counselling for wellness: A holistic model for treatment planning. *Journal of Counselling and Development* 78 (Summer), 251–266.

Nash, D. (2001) On travellers, ethnographers and tourists. *Annals of Tourism Research* 28 (2), 493–495.
Newman, I. and Benz, C.R. (1998) *Qualitative–quantitative Research Methodology: Exploring the Interactive Continuum*. Carbondale, IL: Southern Illinois University Press.
Onwuegbuzie, A.J. and Leech, N.L. (2005) On becoming a pragmatic researcher: The importance of combining quantitative and qualitative research methodologies. *International Journal of Social Research Methodology* 8 (5), 375–387.
Panchal, J.H. (2012) The Asian spa: a study of tourist motivations, 'flow' and the benefits of spa experiences. PhD thesis, James Cook University.
Panchal, J.H. (2013) Tourism, wellness and feeling good: Reviewing and studying Asian spa experiences. In P. Pearce and S. Filep (eds) *Tourist Experience and Fulfilment: Insights from Positive Psychology*. Oxford: Routledge.
Panchal, J.H. (2014) Perceived benefits of spa experiences: Tourists insights from the blogosphere. *International Journal of Tourism Sciences* 14 (3), 24–69.
Panchal, J.H. and Pearce, P.L. (2011) Health motives and the Travel Career Pattern (TCP) model. *Asian Journal of Tourism and Hospitality Research* 5 (1), 32–44.
Pearce, P.L. (1982) *The Social Psychology of Tourist Behaviour*. Oxford: Pergamon Press
Pearce, P.L. (2004) Theoretical innovation in Asia Pacific tourism research. *Asia Pacific Journal of Tourism Research* 9 (1): 57–70.
Pearce, P.L. (2005) *Tourist Behaviour: Themes and Conceptual Schemes*. Clevedon: Channel View Publications.
Pearce, P.L. (2007) The relationship between positive psychology and tourist behaviour studies. Paper presented at the International Academy for the Study of Tourism Conference.
Pearce, P.L. and Filep, S. (2013) *Tourist Experience and Fulfilment: Insights from Positive Psychology*. Oxford: Routledge
Pearce, P.L. and Lee, U.-I. (2005) Developing the travel career approach to tourist motivation. *Journal of Travel Research* 43, 226–237.
Phillimore, J. and Goodson, L. (2004) Progress in qualitative research in tourism: Epistemology, ontology and methodology. In J. Phillimore and L. Goodson (eds) *Qualitative Research in Tourism: Ontologies, Epistemologies and Methodologies* (pp. 3–29). London: Routledge.
Rojek, C. and Urry, J. (1997) Transformations of travel and theory. In C. Rojek and J. Urry (eds) *Touring Cultures: Transformations of Travel and Theory* (pp. 1–19). London: Routledge.
Rossman, G.B. and Wilson, B.L. (1985) Numbers and words: Combining quantitative and qualitative methods in a single large-scale evaluation study. *Evaluation Review* 9, 627–643.
Ryan, C. (2000) Tourist experiences, phenomenographic analysis, post-positivism and neural network software. *International Journal of Tourism Research* 2 (2), 119–131.
Ryan, R.M. and Deci, E.L. (2001) On happiness and human potential: A review of research on hedonic and eudaimonic well-being. In S. Fiske (ed.) *Annual Review of Psychology* (Vol. 52, pp. 141–166). Palo Alto, CA: Annual Reviews, Inc.
Ryan, R.M. and Deci, E.L. (2008) Hedonia, eudaimonia and well-being: An introduction. *Journal of Happiness Studies* 9, 1–11.
Saracci R. (1997) The World Health Organisation needs to reconsider its definition of health. *BMJ (Clinical Research ed.)* 314 (7091), 1409–1410. https://doi.org/10.1136/bmj.314.7091.1409
Sarantakos, S. (1998) *Social Research* (2nd edn). South Yarra, Australia: Macmillan Education Australia.
Seligman, M.E.P. and Csikszentmihalyi, M. (2000) Positive psychology: An introduction. *American Psychologist* 55 (1), 5–14.

Sieber, S.D. (1973) The integration of fieldwork and survey methods. *American Journal of Sociology* 73, 1335–1359.
Smith, J.K. (1983) Quantitative versus qualitative research: An attempt to clarify the issue. *Educational Researcher* 12, 6–13.
Smith, J.K. and Heshusius, L. (1986) Closing down the conversation: The end of the quantitative–qualitative debate among educational inquirers. *Educational Researcher* 15, 4–13.
Smith, M. and Kelly, C. (2006) Wellness Tourism. [Editorial]. *Tourism Recreation Research* 31 (1), 1–4.
Stockmann, N. (1983) *Antipositivistic Theories of the Sciences: Critical Rationalism, Critical Theory and Scientific Realism*. Dordrecht: Reidel.
Tashakkori, A. and Teddlie, C. (1998) *Mixed Methodology: Combining Qualitative and Quantitative Approaches*. Thousand Oaks, CA: Sage.
Tashakkori, A. and Teddlie, C. (eds) (2003) *Handbook of Mixed Methods in Social and Behavioural Research*. London: SAGE Publications.
Teddlie, C. and Tashakkori, A. (2003) Major issues and controversies in the use of mixed methods in the social and behavioral sciences. In A. Tashakkori and C. Teddlie (eds) *Handbook of Mixed Methods in Social and Behavioral Research* (pp. 3–50). London: SAGE Publication.
Teddlie, C. and Tashakkori, A. (2012) Common 'core' characteristics of mixed methods research: A review of critical issues and call for greater convergence. *American Behavioural Scientist* 56 (6), 774–788.
Tribe, J., Dann, G. and Jamal, T. (2015) Paradigms in tourism research: A trialogue, *Tourism Recreation Research* 40 (1), 28–47.
Vidich, A.J. and Shapiro, G. (1955) A comparison of participant observation and survey data. *American Sociological Review* 20, 28–33.
Walle, A.H. (1997) Quantitative versus qualitative tourism research. *Annals of Tourism Research* 24 (3), 524–536.

14 Paradigms: A Supervisor's Perspective and Advice

Philip L. Pearce

Introduction

In the last two decades a popular, branded set of T-shirt have been sold throughout the western world. The image of a crown appears on top of the text with the lead theme **Keep Calm and Relax.** I want to reuse this signature meme and suggest immediately to postgraduate students: **Keep Calm and Relax** *about paradigms*. The topic need not pose a daunting dilemma or be a consuming concern. Indeed, researchers at any level do not have to agonise over a single, agreed-on frame in which to locate their work. Nevertheless, I will argue that it is valuable to reflect on your assumptions, to proceed mindfully in conducting studies and to have respect for other styles of work, but there need not be a life-long commitment to a particular paradigm. Indeed, it is not obligatory that a tourism thesis is entirely consistent with one paradigmatic style if different kinds of questions are being asked. Instead, I will assert that the whole constellation of the research piece revolves around the aims and purposes of the project. In this chapter I will review, from a supervisor's perspective, the topic of paradigms and the guiding styles for doing good work in tourism. My perspective derives from the experience of reading germane literature, of being given advice and providing advice, and seeing how that counsel has been evaluated by examiners of theses and journal reviewers.

The contents of the chapter consist of four main sections. In the first part of the chapter, I document the rise of key expressions and note the changes in the use of pivotal terms which accompany the broad treatment of the term paradigm. Next, I will characterise the fundamental positions. Pockets of controversy are included in this second part of the review. In a third set of comments, I will dismiss claims for new paradigms as the ways to do research. This rejection of new paradigmatic candidates consists of considerable value in the themes and less well researched topics. Nevertheless, I see the collation of ideas on such themes as mobility, and Asian worldviews as not truly offering systematic new ways of being a researcher. The fourth part of the review builds a more personal narrative. I hope that by the time you are reading this chapter, I will have successfully supervised to completion close to 50 PhD students. Together with the supervision of Honours and Masters students, my total portfolio of mentoring emerging scholars amounts to around 100 research theses candidates. It is a deep well from which to draw examples and relate tales of experience. Further, like all researchers, I have experienced personal moments of truth, of confrontation and of reconciliation in the work I have done. A little of this personal story, as well as the stories of supervising many junior colleagues, will be employed to offer succour and inspiration to those who follow the research pathway in tourism academic life.

A Special Language

We can begin with the picture of a North American graduate student, who has recently arrived in Australia. On their first visit to their supervisor's home, they enjoy the food and plenty of good Australian wine. The need to relieve their well-filled bladder descends on them. Politely they ask for the bathroom. On finding the designated room, they are somewhat perplexed – there is no facility in this room to do what they need to do. Returning to the assembled gathering, this time they ask for the washroom. Directed to a room containing only a washing machine and clothes dryer, their frustration mounts. They return quickly and mutter more directly that they need to urinate. At last, they are pointed in the direction of the toilet and order is restored. The implication here is that language and the way we use it, is sensitive to cross-cultural influences, it changes over time, and further, imprecisely used terms can confuse everyone.

The concept of paradigms follows this pattern of changing and contextually dependent use. It is a concept which now has interpretations which do not reflect its original scientific and scholarly use. Politicians now talk about changing the economic paradigm, football coaches refer to new playing strategies in terms of fresh paradigms and the police express enthusiasm for the new paradigm of community-centred surveillance. By way of contrast, the academic meaning of paradigm was born in the writings about the history of science. It was employed to help understand the revolutions in thinking which followed transformative discoveries. While the term has both European and North American roots (Dann, 1997), it is somewhat obligatory to turn to the American Thomas Kuhn to extract a foundation definition (Tribe et al., 2015). Kuhn wrote in his 1962 book *The Structure of Scientific Revolutions* that the concept:

> stands for the entire constellation of beliefs, values, techniques and so on shared by members of a given community. (1962: 175)

It is a big concept. And rather importantly, the emphasis in Kuhn's analysis is that these agreed on ways of doing things in science are controlled and policed. It is perhaps appropriate to invoke the spirit of Foucault here and his Panopticon model: the idea that the inmates of a prison internalise the gaze of their guards (Foucault, 1980). So too with paradigms. There are gatekeepers and control agents. These are the journal editors, supervisors, promotion boards, heads of departments and academic societies, who all require those under their influence to behave in accord with the rules of the system. But the inmates also survey one another and serve to institutionalise the ways to conduct appropriate research.

Kuhn's concept needs to be understood by referring to the context in which it was first used and its purpose. It was not primarily about disciplines. It was not about theories (they are a part of paradigms). It was about the whole world of science, and the long reach of its history since the Renaissance. The frame of reference was specifically about studies of the physical and natural world. Kuhn at first did not think his concept was appropriate to social sciences (Harrison, 2017). Later he changed his mind somewhat and cautiously applied his views to groups who share a common language, which might be taken to mean specific discipline or study fields.

Thomas Kuhn lived for quite a long time (1922–1996). He contributed ideas and revisions to his thoughts some 30 years after he first extolled paradigms as a way to bring clarity to the changes in science. In later years, he had virtually abandoned the term. It appears he was frustrated that many others used it loosely (Walker, 2010). He replaced the emphasis on paradigms with a simpler test of the commonality of research effort. He used the expression 'incommensurability' – the degree to which study areas fail to share common standards of measurement (Harrison, 2017). If research groups do share measurement standards, then there is most likely going to be a strong in-group who seek validation from each other only. If no such standards of agreement exist, then there is no de facto grouping; that is, no paradigmatic claim can be made for such a group.

In his 1990 revisions, Kuhn placed considerable stress on this ability of a group of researchers to understand methods and results. Many textbook analyses of paradigms refer more directly to four sources for defining specific paradigms – the key terms being ontology, epistemology, axiology and then also methodology (Jennings, 2010). Using these terms, the following dominant 'constellation of beliefs, values, techniques and so on shared by members of a given community' to re-use Kuhn's phrase, appear in the literature: positivism and post-positivism, constructivism, and critical theory and its variants. These are sometimes called the master paradigms (Tribe et al., 2015), though it needs to be pointed out quite quickly that the names and labels are not uniform across the diverse areas of social science (Outhwaite, 2000).

In accord with good educational practice, it is possible to link our exposition of these terms with Bloom's taxonomy for comprehending new material (Bloom, 1956; Krathwohl, 2002). It is desirable in understanding a new set of issues to firstly define and describe (comprehend), then model and analyse the topic (explain), and finally have the ability to apply, evaluate and synthesise (create) your position (Marzano & Kendall, 2006). The following succinct presentation of the key terms follows this trajectory.

Ontology is our view of reality. For example, the questions to be asked here are those of 'are such entities as buildings, Australia, and fair play all real?' That is, we are asking is there something out there separate from ourselves or is there no objective reality, just a constructed and assumed world? Potential ontological positions include being a naïve realist (yes there are 'buildings' and 'Australia', but maybe not 'fair play'), a critical realist (yes there is a possible reality of 'fair play' but we have to discuss and share a lot of views to confirm that the concept is sound), a position of relativism (there is not really any objective reality for any of these terms, it depends on who you are and your own point of view) or a pragmatist stance (there is a reality if we all say so; what 'works' is what is real or true).

The second term, epistemology, focuses on how we know what we say we know. A first position is that we adopt a representational epistemology. This means we construct a summary device (a pithy saying, a model, a formula, or a symbolic image) of an objective reality. By way of contrast, a second position – that of a transactional or subjectivist epistemology – asserts that we cannot separate ourselves from what we know and who we are. Instead, we have to negotiate and argue for our understanding, for example through the use of logic, by exploring many examples, participating in discussion, seeking group consensus and negotiating interpretations.

In the textbook analyses, paradigms are also built on dominant methodologies and methods (Jennings, 2010). These terms differ in that methodology refers to the logistics, the whole design of the study, while methods are the tools employed to collect data. One further term completes the set required to build the master paradigms discussed in much tourism study. Axiology is employed as a further defining feature – essentially the expression captures the philosophical study of goodness. It is employed to emphasise the value, in the widest sense, of the research and draws attention to the likely beneficiaries of studies.

Fundamental Positions

Taken together, these components result in positivism, post-positivism, constructivism and critical theory. Positivism is a constellation tied to the physical sciences and pertains to those study fields where the objects of the research are affected minimally by the presence and actions of the observer. Some tourism economists are effectively positivists, most noticeably when their data are pre-existing statistical records and policy documents. True positivists in tourism are naïve realists, they employ a representational epistemology, use a suite of quantitative analyses and offer a service to the business of tourism.

For the post-positivists, the ontology is largely one of critical realism – there is a world external to the observer, but the researcher has to do some work to negotiate and check on its existence. The epistemological view is that researcher objectivity and distance is not possible, but there are better arguments than others for assuring the development of knowledge. If a community of scholars or others can agree on a concept, then it can be represented. This is a subjectivist epistemology, although the extent to which this is embraced by post-positivists varies. In the research process, manipulation checks for experiments, and the use of pilot surveys and emic perspectives in constructing surveys, represent the softer dimensions of the post-positivist position. Quantitative methodologies prevail. The axiology is largely instrumental, the work can be seen as serving the good of the tourism management system and often businesses within that system.

The problems with some post-positivist assumptions and measures have resulted in the growth and now strong cohort of those described as constructivists. The reactivity problem has been of particular concern to constructivists. For example, the topic of tourists' sexual behaviour on holidays and potential disease transmission resulting from these actions is important from the perspective of public health as well as individual well-being. Pisani (2010) has observed that who asks the questions about sexual behaviour and how they are asked has a massive influence on the accuracy of the responses. Indeed, interpretivists argue that knowledge can never be certified as true, and sustained interaction is the only way to access people's constructions of their world. The ontological stance for constructivists is relativist. Getting close to understanding what is happening in people's lives involves close consultation and is constructed among people. The epistemology is transactional; as researchers we cannot separate ourselves from who we are and what we know. At the very least, we need to state and declare these personal influences in reporting our studies. Constructivists favour accessing people in natural settings and conversing with them to access information. Conducting interviews and focus groups, reading people's narratives/stories, and analysing sections of talk as well as online records, are all used and valued (cf. Wu & Pearce, 2014). For axiology, moral and power questions are important, and the open expression of researchers' values are encouraged.

Arguably, the most controversial paradigm or constellation is defined by the expression critical theory. It is also known as the 'critical turn'. The puzzle for many is whether the writing done from within this group is really research, possibly instead it is forceful advocacy for those who need a public voice. Again, while there is marked internal variability among those who might label themselves as critical theorists, the expression broadly characterises a position where researchers question the purpose of what they do and ask whose interests are served by their studies (Ateljevic *et al.*, 2013). A strong axiological position, a concern for the welfare of others, typically those who are marginalised, underpins this paradigm. The value in terms of axiology is clear – authors who are in this group seek change and an improved world. Critical theorists have much in common with applied sociologists and philosophers who focus on the use of language (Grayling, 2006; Harris, 2005). Ascribed labels and identity groupings can all perpetuate power and privilege. Further, critical theorists are not content with simply describing tourism settings and situations, they are often actually trying to induce change.

Critical theorists can certainly be conceived as embracing a broad church. There is effectively a family of approaches with similar goals and linked concerns. Some of the intentions of feminist theory, and more recently that of queer theory, offer similar voices about those who have been marginalised (Waitt & Markwell, 2014). Such studies question the roles and labels for women and men and reveal the dominant powerful majorities in shaping acceptable public conduct. Those who are concerned with the lingering hands of colonialism and imperialism in the global South – effectively the world's poorer countries – are close cousins of the critical theorists (Wijesinghe *et al.*, 2019). There are also links to action research, notably as it is used in social work and education. This style of study follows an iterative approach of participating in change through research being immediately implemented (Kemmis *et al.*, 2013).

For critical theorists, reality has been created by past practices, notably political, social, cultural and economic forces which have shaped the incumbent social structures. In an ontological sense, they are never naïve realists and are opposed to those who take the observed status quo as inevitable and necessary. Critical theorists suggest that treating current situations and practices as real is limiting, and at times they need to be re-examined. They are at least critical realists and quite often relativists in their ontological stance. An example about wrongful imprisonment can illustrate the position. MacFarlane and Stratton (2016), together with many others, point to the high rates of imprisonment among indigenous people in Canada and Australia. They argue that wrongful conviction may be 'a function of a managerialist approach to criminal justice prioritising efficiency, expediency and risk management over due process' (2016: 303). By adopting a relativist view, critical theorists can prompt a re-examination of some of the evidence offered in these trials. For example, the reasons indigenous women plead guilty when they are not, may be determined by fears of domestic violence (Roach, 2015). A relativist perspective is therefore helpful in adopting another view and challenging the established legal determination.

Further, for critical theorists a transactional or subjectivist epistemology prevails. Individuals are embedded in their constructed realities. There is, though, a key issue here. The problem for researchers using this approach is the likelihood that they will only find what they are looking for; searching with their personal torch rather than illuminating an entire issue through examining a broad spectrum of possibilities. Fostering conversation and reflection, encouraging dialogue to understand conflict and tension and questioning people's current experience with regard to values are the methods and methodological styles of critical theorists. In particular, an acute awareness of the researcher's positionality and role in the research process requires close reflection and personal attention. In this analysis, the researcher's biases and values need to be outlined to assist the reader in determining the author's orientation to the topic (Tong et al., 2007). If these processes are followed thoroughly, then the limitations of looking for and finding only what one already thinks can be lessened.

Critical theorists and even constructivists have attracted some sharp and sometimes cynical attention. Joseph Mazanec, the respected Austrian tourism market researcher who has influenced many scholars under his tutelage, advises that youthful tourism researchers should study mathematics and computer science rather than muddling along as narrative scientists. He suggests that to follow the latter is to risk being torn apart (Mazanec, 2011). In the United States, Alan Sokal, Professor of Physics at New York University, parodied the writing of critical theorists and constructivists and had his hoax piece published in a leading sociological journal, *Social Text* (Sokal, 1996). Gould (2004) called it a sorry affair that simply alienated scholars from different traditions, but the Sokal hoax illustrated at least one point of importance. Fashionable research directions need to be subject to calm logical analysis rather than enthusiastic (and presumably tiring) jumping on and off bandwagons.

The original Sokal hoax has been overshadowed recently by its sequel. Referred to as Sokal squared, three authors – magazine editor Helen Pluckrose, mathematician James Lindsay and philosopher Peter Boghossian – worked together to write 20 fake articles (Beauchamp, 2018). Seven were accepted by journals in humanities and critical studies. The studies were poorly constructed in terms of basic methods and research protocols and the content was sometimes blatantly absurd. Some of the more extreme examples included supposedly thousands of observations of dog rape in a public park, a feminist rewrite of *Mein Kampf*, and the role of sex toys in sensitising men to homosexual acts. The point of the effort, the authors argued, was to highlight that even extreme and somewhat ridiculous topics could be accepted for publication if the right language was used and the values portrayed in the work matched what they termed the prevailing ethos of grievance studies where the group being described was apparently wronged or misunderstood (Grech, 2019; Kafka, 2018). Pluckrose, Lindsay and Boghossian have outlined their views in a long and revealing YouTube interview with the globally influential psychologist Jordan Peterson. The authors who have been strongly criticised, even vilified, for abusing the academic reviewing system with time-consuming nonsense, assert that politically fashionable topics should not be accepted if rigorous scholarship is not involved (Peterson, 2019).

Perhaps not surprisingly, all the published articles have been withdrawn from circulation by the various editors.

New Paradigm Candidates

At least within tourism and hospitality studies, there are several suggestions that the paradigms and ways of representing commonalities in the research effort are being refreshed (Tribe *et al.*, 2015). The most prominent candidate with a journal of the same name is that of Mobilities, or as some advocates position the work, the NMP – the new mobilities paradigm (Coles, 2015; Urry, 2007). The claims prompt the question – what would a constellation of beliefs and approaches have to do to constitute a new enclave of self-approving scholars? The answer is a lot. It would have to develop new ways of approaching at least some novel topics with fresh tools and be capable of being assessed only by its own tight circle of adherents. In several reviews of this approach, scholarly opinion recognises a contribution from the way the work in mobilities stresses the conjunction of moving related phenomena – ideas, people, bodies, events and some places. For example, Cohen and Cohen (2015) who are somewhat enthusiastic about the new mobilities approach and its ways of looking at key topics, nevertheless lament that it offers no new tools or research approaches. Faulconbridge and Hui (2016) regard the movement as decidedly open and ready to include anyone who wants to join – a position at odds with the basic premises about paradigm requirements for incommensurability and control. In brief, it can be stated that the mobilities approach does not meet the necessary performance indicators (cf. Harrison, 2017). Some of the topics studied may be fresh but that alone does not command attention as a new way of doing social science.

There is perhaps an accompanying and slightly cynical view here that remembers that one of the leading scholars of mobilities, the late John Urry, also shaped the study of tourism with an influential book on tourist gaze (Urry, 1990; Urry & Larsen, 2011). In time he moved on from that commanding work to embrace embodiment and a richer awareness of the sensual world of the tourist. Mobilities was his next passion. In adopting these molar concerns, there is the recurring danger that other big topics are still missing from the single-minded formulations. Historical, political, economic and spatial aspects of the social world must surely also be candidates for focused interest, but that does not mean that they can or should be studied in a unique way and be new paradigms.

A rather different kind of claim for a new paradigm arises from what was once called the East and to a lesser extent the South. The very language in that sentence is indicative of the motives for this paradigm push. It is clear that tourism study was first formulated in Britain, Europe and the United States, closely followed by Australia, New Zealand and

Canada. English-speaking countries persist today as sources of researcher productivity in tourism. European scholars also contribute to English outlets as well as writing in their own languages. As Morris (2011) notes in his book, *Why the West rules for now*, the rise of Asian economies, as well as some improving living standards in poorer nations everywhere, raises questions of the sharing of global opportunity and understanding. Such power shifts include shifts in the research world. Scholars from the East, especially China, South Korea, Malaysia and to some extent India, are now contributing repeatedly to the tourism journals. There are others too, often key individuals from Europe, the Middle East, Africa and South America who are serving the role of creating local tourism research communities of international note. And one facet of this contribution is the possibility that researchers from these locations might have new ways to think about the phenomenon of tourism. For some, the control of the tourism agenda in research provides echoes of a colonial past. In other countries which were never fully colonised, there is an assertiveness in wanting the voices of their communities heard and their concerns conceptualised in ways meaningful to their citizens.

But here too, there is no reason to believe that a new paradigm has emerged or indeed is necessary. Fresh voices, new forms of interpersonal relations, varied customs and values, different styles of tourism, compelling local needs for tourism to have a positive effect on local people, are all of international as well as local interest, but they do not constitute a new constellation of beliefs and methods for the research community. Arguably, we already have enough variability on the patterns and styles of research to take on these challenges in tourism study. Instead of a hunt for new paradigms, a much sharper reality confronts tourism scholars from emerging destinations. Despite good intentions to conduct developing country tourism research, there are not often the funds to generate the activities which could contribute to a flourishing and ethical local research agenda (Oktadiana & Pearce, 2017).

Personal Encounters

I first encountered the then murky word of the philosophy of science on my second day in Oxford. As an Australian I had come to study for my PhD in psychology. Standing in the morning tea line, a senior bespectacled gentleman asked me what was my ontological position for my planned work in tourism? Befuddled by the question, I made a stab at an answer, replying that tourists were becoming a 'real issue'. This answer appeared to greatly amuse the incumbent Professor of Moral Philosophy, who began muttering strangely about colonials and Australian pragmatism. I later learned that pragmatism was actually a philosophical school in my home country; indeed a platform which attempted to avoid language tricks in understanding key philosophical positions.

The previous little account is worth recalling because, as was the case in my own lack of readiness to study a higher degree, many new postgraduate researchers are not conversant in the terminology and philosophy of research. Courses to that effect have started to be included in research degrees but supervisors still play a key role in interpreting the dry material for a specific project. It is my candid view that for the first 15 PhD people I supervised until the end of the twentieth century, direct discussions of paradigms did not occur. The work supervised was indeed diverse and crossed a number of boundaries and perspectives, which in more contemporary times could be seen as mixed methods and blended versions of post-positivism and constructivism. I can elaborate on the topics to indicate some of the fine work that diligent doctoral candidates produced. There were PhDs built on field experiments about helping strangers in small towns versus cities, ethological observational work on the play and leisure behaviour of children, large-scale survey work on community health and travel, testing of scenarios for the re-use of heritage buildings for tourism, and quantitative appraisals of the ratio of travel agents to consumers in cities linked to their likely business success. Further studies with a more direct tourist behaviour focus included assessing cultural tourists' interests in artefacts and museums in Malaysia, tracking backpackers' decision-making during their Australian trips, noting motivational drivers' differences in the emerging Korean market and pioneering the assessment of online booking for bed and breakfast accommodation.

And yet, despite this diversity and the use of examiners varying in their backgrounds from psychology, tourism studies, geography, sociology and anthropology, there were no specific comments I can recall where the need to document the paradigm framing the work was mentioned. There were plenty of plaudits for ensuring the aims were clear and met, that the methods employed were used correctly, and that care was taken in explaining sampling. There were remarks about valuing the researcher's cautious awareness of the limitations of their own position and approach. But not any insistent demand for paradigm analysis and discussion.

Within a decade that had changed. As my students submitted theses in the 2000–2010 period, and particularly when those theses went to sociologists and tourism researchers in Britain, some new feedback emerged. I remember distinctly a thorough piece of work done on Asian seasonality. It was indeed a rather positivist approach which employed existing monthly figures and noted a series of patterns unlike those seen in Europe. It won a Best Paper award at a major Asian conference. A shortened version of the work was published in *Tourism Economics*. Nevertheless, the examiners asked the student to insert a systematic review of the paradigms of research before they were willing to pass the thesis. At the same time, I started to see such renditions about paradigms appearing as almost standard practice in the theses I marked from the UK, Australia and New Zealand. Clearly a shift in the philosophical awareness about tourism researchers' approach to their own work had emerged (cf. Tribe, 2001, 2009). By the middle of the decade, I was advising my own students to reflect more explicitly on these issues and write at least truncated statements about their awareness of mixing methods, identifying their own committed interests, and however loosely, describing their allegiance to the major or master paradigms.

The Current Situation

Since about 2005, my PhD students have mostly presented a tour of the paradigms-related literature. The approach has been effective in avoiding negative reviews and is now generally glossed over by examiners as an apparently necessary (and easy) box to check off in their review. This might all seem to be a happy state of affairs – students knowing more about their assumptions and proceeding to fit into the required dictates of certain constellations and systems of doing good work. The only difficulties which seem to persist have not come from thesis work. Instead, in day-to-day intra university academic commentary, in reviews, and in conferences, it can be observed that there are some hardened positions which can confront the PhD researcher (and indeed their supervisors). One such position is what the social science philosopher Rom Harre once described as belonging to the nasty little positivists (cf. Harre, 2000). Objections to research efforts from this group focus on sampling issues, the ability or inability to measure constructs clearly, the push to use more elaborate statistics, a belief that the best research must produce models and formulaic accounts of the world and a disdain for case study efforts as well as a lack of enthusiasm for most interpretive and critical theory work. For supervisors, there is a need to identify these individuals and not ask them to examine the more expansive kinds of work. At conferences and in reviews, they may be people to manoeuvre around unless there is a readiness for a stand-up academic brawl.

Unreconstructed positivists are rare, but another cohort is now more common; that of the confirmed critical theorists who sometimes align with constructivists in challenging the purpose and relevance of rather bland, empirical and industry-oriented studies. The perspective is crystallised by recent editorial advice from Chris Ryan, former editor of *Tourism Management*; he offers the following advice to those seeking to submit papers:

> In the case of manuscripts based solely on a quantitative analysis using data derived from a single survey there is a strong preference that such papers be accompanied by supplementary materials such as interview material, data from another survey to permit comparative or longitudinal analysis, or similar supplementary data sets to provide further insights into the quantitative data analysis. This also applies to panel data. (Guide for authors, https://www.journals.elsevier.com/tourism-management)

There is an associated warning with this edict that papers based on, for example, structural equation modelling alone, will not be accepted. The strong position reflects the tip of the demands from the cohort of critical theorists who hold clear views that tourism researchers should pursue topics which offer community change and boost empowerment (Caton, 2012; Ren et al., 2010; Wilson et al., 2008).

Conclusion

For the graduate student who may be presenting a specific kind of study to audiences where the constellation and systems of researcher beliefs differ from their own, I can offer a detour around messy conflicts. The systems of study which prevail in tourism function for different questions and separate kinds of research aims. In effect, the universal reply to many of these aggressive paradigm-like confronting remarks is: 'That is not relevant to the question I am addressing'. In using this defence, it is mandatory to have a very clear question and attendant precise and firmly articulated aims. The methods and philosophical positions in terms of ontology, epistemology and axiology all flow from these very clear research aims. There can, of course, be a debate about whether the questions and aims selected are the best ones for that topic – this is a consideration which links to what Corley and Gioia (2011: 23) called pre-science; a careful review of the topic involving 'discerning what we need to know and influencing the intellectual framing of what we need to know to enlighten both academic and reflective practitioner domains'. But the standby response of 'that is not relevant to the question I am addressing' is a useful and direct response, not a subterfuge or a devious escape from academic harassment. As a supervisor, I want to see work done well to meet clear questions and specific aims. Paradigm wars are a distraction to that goal.

References

Ateljevic, I., Morgan, N. and Pritchard, A. (eds) (2013) *The Critical Turn in Tourism Studies: Creating an Academy of Hope* (Vol. 22). New York: Routledge.

Beauchamp, Z. (2018) The controversy around hoax studies in critical theory, explained. See https://www.vox.com/2018/10/15/17951492/grievance-studies-sokal-squared-hoax (accessed 22 May 2019).

Bloom, B.S. (1956) *Taxonomy of Educational Objectives: The Classification Of Educational Goals: Cognitive Domain*. London: Longman.

Caton, K. (2012) Taking the moral turn in tourism studies. *Annals of Tourism Research* 39 (4), 1906–1928.

Cohen, E. and Cohen, S. (2015) Beyond ethnocentrism in tourism: A paradigm shift to mobilities. *Tourism Recreation Research* 40 (2), 157–168.

Coles, T. (2015) Tourism mobilities: Still a current issue? *Current Issues in Tourism* 18 (1), 62–67.

Corley, K.G. and Gioia, D.A. (2011) Building theory about theory building; What constitutes a theoretical contribution. *Academy of Management Review* 36 (1), 12–32.

Dann, G. (1997) Paradigms in tourism research. *Annals of Tourism Research* 24 (2), 472–474.

Faulconbridge, J. and Hui, A. (2016) Traces of a mobile field: Ten years of mobilities research. *Mobilities* 11 (1), 1–14.

Foucault, M. (1980) *Power/knowledge: Selected Interviews and Other Writings, 1972–1977*. London: Pantheon.

Gould, S.J. (2004) *The Hedgehog, The Fox and The Magister's Pox: Mending and Minding the Misconceived Gap Between Science and the Humanities*. London: Vintage.

Grayling, A.C. (2006) *The Heart of Things. Applying Philosophy to the 21st Century*. London: Phoenix.

Grech, V. (2019) Write a Scientific Paper (WASP): Academic hoax and fraud. *Early Human Development* 129, 87–89.

Harré, R. (2000) Varieties of theorizing and the project of psychology. *Theory & Psychology* 10 (1), 57–62.

Harris, D. (2005) *Key Concepts in Leisure Studies*. London: Sage.

Harrison, D. (2017) Tourists, mobilities and paradigms. *Tourism Management* 63, 329–337.

Jennings, G. (2010) *Tourism Research* (2nd edn). Milton: John Wiley & Sons.

Kafka, A.C. (October 3, 2018) 'Sokal squared': Is huge publishing hoax 'hilarious and delightful' or an ugly example of dishonesty and bad faith? *The Chronicle of Higher Education*. Retrieved from: https://www.chronicle.com/article/sokal-squared-is-huge-publishing-hoax-hilarious-and-delightful-or-an-ugly-example-of-dishonesty-and-bad-faith/

Kemmis, S., McTaggart, R. and Nixon, R. (2013) *The Action Research Planner: Doing Critical Participatory Action Research*. Amsterdam: Springer Science & Business Media.

Krathwohl, D.R. (2002) A revision of Bloom's taxonomy: An overview. *Theory Into Practice* 41 (4), 212–218.

Kuhn, T. (1962) *The Structure of Scientific Revolutions*. Chicago: The University of Chicago Press.

Kuhn, T. (2000) *The Road Since Structure. Philosophical Essays 1970–1993 with an Autobiographical Interview*. Chicago: University of Chicago Press. (published after Kuhn's death)

MacFarlane, J. and Stratton, G. (2016) Marginalisation, managerialism and wrongful conviction in Australia. *Current Issues in Criminal Justice* 27 (3), 303–321.

Marzano, R.J. and Kendall, J.S. (eds) (2006) *The New Taxonomy of Educational Objectives*. New York: Corwin Press.

Mazanec, J. (2011) Marketing science perspectives of tourism. In P.L. Pearce (ed.) *The Study of Tourism Foundations from Psychology* (pp. 79–92). Bingley: Emerald.

Morris, I. (2011) *Why the West Rules for Now*. London: Profile Books.

Oktadiana, H. and Pearce, P.L. (2017) The 'bule' paradox in Indonesian tourism research: Issues and prospects. *Asia Pacific Journal of Tourism Research* 22 (11), 1099–1109.

Outhwaite, W. (2000) The philosophy of social science. In B.S. Turner (ed.) *The Blackwell Companion to Social Theory* (2nd edn, pp. 47–70). Oxford: Blackwell.

Peterson, J. (2019) Interview with grievance studies hoaxers. See https://www.youtube.com/results?search_query=Jordan+Peterson++sokal+hoax

Pisani, E. (2010) *The Wisdom of Whores: Bureaucrats, Brothels and the Business of AIDS*. London: Granta Books.

Ren, C., Pritchard, A. and Morgan, N. (2010) Constructing tourism research: A critical inquiry. *Annals of Tourism Research* 37 (4), 885–904.

Roach, K. (2015) The wrongful conviction of Indigenous people in Australia and Canada. *Flinders LJ* 17, 203.

Sokal, A.D. (1996) Transgressing the boundaries: Toward a transformative hermeneutics of quantum gravity. *Social Text* (46–47), 217–252.
Tong, A., Sainsbury, P. and Craig, J. (2007) Consolidated criteria for reporting qualitative research (COREQ): A 32-item checklist for interviews and focus groups. *International Journal for Quality in Health Care* 19 (6), 349–357.
Tribe, J. (2001) Research paradigms and the tourism curriculum. *Journal of Travel Research* 39 (4), 442–448.
Tribe, J. (ed.) (2009) *Philosophical Issues in Tourism*. Bristol: Channel View Publications.
Tribe, J., Dann, G. and Jamal, T. (2015) Paradigms in tourism research: A trialogue. *Tourism Recreation Research* 40 (1), 28–47.
Urry, J. and Larsen, J. (2011) *The Tourist Gaze 3.0*. London: Sage.
Urry, J. (1990) *The Tourist Gaze: Leisure and Travel in Contemporary Societies*. London: Sage.
Urry, J. (2007) *Mobilities*. Cambridge: Polity.
Waitt, G. and Markwell, K. (2014) *Gay Tourism: Culture and Context*. Abingdon, Oxon: Routledge.
Walker, T.C. (2010) The perils of paradigm mentalities; revisiting Kuhn, Lakatos and Popper. *Perspectives on Politics* 8 (2), 433–451.
Wijesinghe, S.N., Mura, P. and Culala, H.J. (2019) Eurocentrism, capitalism and tourism knowledge. *Tourism Management* 70, 178–187.
Wilson, E., Harris, C. and Small, J. (2008) Furthering critical approaches in tourism and hospitality studies: Perspectives from Australia and New Zealand. *Journal of Hospitality and Tourism Management* 15 (1), 15–18.
Wu, M.Y. and Pearce, P.L. (2014) Appraising netnography: Towards insights about new markets in the digital tourist era. *Current Issues in Tourism* 17 (5), 463–474.

15 Into the Future: Moving Forward with Reflective Practice on Paradigms

Josephine Pryce

Introduction

This book sought to bring to you, the reader and scholar, reflections on the endeavours of PhD candidates and emerging scholars as they engaged with the notions of research paradigms and to comprehend how and why the underlying assumptions of their respective paradigm informed their topic and research design. Each chapter portrayed the authors' challenges and enlightenment as they journeyed to seek clarity in understanding their paradigm and grounding of their research in ways that are true to their paradigm. For the editors of this tome, attention is drawn to Ling and Ling's (2016: 364) observation that, 'paradigms, as well as being organisational and analytical devices used to categorise research undertakings, can be concepts that transform thinking about oneself as a researcher, a teacher or a learner'.

Each of the authors in this book has shown how they are linked to their research through their paradigm and how different paradigms can '[result] in new knowledge, insights, understandings, problematics, unknowns, imaginings, and possibilities' (Ling & Ling, 2016: 351). They have shown the importance of 'being clear about the research paradigm that applies' (Ling & Ling, 2016: 345) and of the underlying assumptions that inform the associated ontology, epistemology and methodology. The chapter authors have also illustrated that regardless of the paradigm one is working from, whether positivism, interpretivism or pragmatism, 'the representation of reality is always interpreted through symbolic representations such as numbers and words' (Merriam & Grenier, 2019: 25).

Undertaking the exercise of engaging with the world of paradigms assists in ensuring the research process is appropriate, rational and structured, and informs the outcomes, allowing for knowledge to be built in defensible ways and further discoveries to be made. Yet, the exercise of engaging with paradigms can be dynamic and illuminating. It can shape and reshape a researcher's focus because as the researcher is more mindful of the various worldviews, they can open their mind to all sorts of possibilities in adopting an ontological, epistemological or methodological stance. There is opportunity for them to combine ontological perspectives (as we saw in Chapter 13) or employ more contemporary epistemological ways (as evidenced in Chapters 9 and 10).

The PhD research process is understood to be a rigorous, systematic, organised investigation where data are collected, collated, analysed and interpreted, and the results disseminated. It seeks to, 'understand, describe, predict or control an educational or psychological phenomenon or to empower individuals in such contexts' (Mertens, 2005: 2). In so doing, the researchers' own theoretical framework or 'way of seeing the world', i.e. paradigm, influences the research. The emphasis given to discussions of paradigms during the PhD journey is varied but the influence of one's paradigm is powerful and often underestimated. It 'sets down the intent, motivation and expectations for the research' (Mackenzie & Knipe, 2006: 2). Hence, spending some time reading, thinking, discussing and reflecting on one's paradigm is important and should be essential for students undertaking a Doctor of Philosophy. Wider engagement with discussions of paradigms, ontologies, epistemologies and methodologies, can enrich and strengthen one's research. We have seen from the authors that the scholarship that is generated by knowledge supposedly bound by paradigmatic assumptions need not be constraining but rather it can be liberating and lead to deeper and richer perspectives and discoveries that can inform policy and practice. Understanding the importance of exploring paradigms to discover one's own 'frame of reference' is empowering for the researcher. In this final chapter, we set out to build on the insights from the previous chapters and to extend the scholarship of paradigms.

Paving the Way for Multidisciplinary and Interdisciplinary Research

These days, it is recognised that the concept of 'academic tribes', i.e. where researchers are constrained to the paradigms of their respective disciplines, is giving way to 'fields of study' that allow for knowledge to be expanded and created in ways that are not necessarily free from the researcher's basic values, assumptions and interests (Tribe, 2004). Rather, such multidisciplinary and interdisciplinary approaches allow for powerful construction of knowledge.

Awareness of paradigms allows for meaningful engagement with researchers from other paradigms and opportunity to produce research that is enriched by the insights of individuals from various paradigms. This approach allows researchers from a plurality of philosophical positions to produce research that makes a difference to people's lives and the world that we live in. For example, applied clinical research in health is increasingly informed by both scientific and social research. In health, researchers from varying paradigms are already collaborating in multidisciplinary and interdisciplinary research. Allsop (2012) illustrates how public health research is reliant on collaborative approaches. She presents the example of 'epidemiological research that investigates the causes of ill health, health economics that applies economic models to assess the costs and benefits of . . . health care, and [its] translational research that converts research findings into products to be used in the treatment of patients' (Allsop, 2012: 18). Such research is informed by both positivist and interpretivist 'ways of knowing' and 'ways of doing' and relies on the weaving of the two to 'provide a framework' that respectfully accommodates both paradigms.

Within this research there is no room for the 'paradigm wars' that Hammersley (1992) referred to and Professor Pearce made mention of in Chapter 14. Hammersley (1992: 141) notes some people may think that conversations around paradigms should be 'left to the philosophers' and counters this view by referring to the work of Guba (1990) and Smith (1989) and their contention about 'methodological holism [as] the idea that researchers' philosophical and practical assumptions form, or should form, a coherent whole with the first determining the second' (Smith, 1989: 142). Hammersley (1992: 142) does not agree entirely with this argument and holds that philosophical assumptions are not 'a privileged starting point' because in the process of conducting and reflecting on our research, we can begin to question our philosophical assumptions, resulting in shifts in one's research practices. He concludes that the war between paradigms is likely to go on. Bryman (2006), however, concedes that peace has now settled on the paradigm wars. He argues that the emergence of mixed methods and the evolution of pragmatism have shown that quantitative and qualitative research is compatible and that the marriage of the two approaches can lead to quality research that transcends any ontological and/or epistemological divides. Adding to the conversations on mixed methods, Allsop (2012: 29) states that 'one type of method is not superior to another . . . [but rather], each serves a particular purpose' and allows the researcher to engage in the critical thinking of the knowledge produced by the respective methods utilised.

Shannon-Baker (2016: 321) emphasises that, 'Paradigms are not static, unchanging entities that restrict all aspects of the research process. Instead, paradigms [are] . . . a guide that the researcher can use to ground their research'. The researcher uses the paradigm to design the research process based on their beliefs. Parallel research on the same research problem utilising a different paradigm can yield other recommendations that can equally make valuable contributions. We see this complementarity of paradigms in mixed-methods research, as discussed in Chapter 13 of this book. This highlights that each paradigm can make valuable contributions to the collective research and enhance research practice. The result is an integrated approach that enables understanding of and insight into what is the 'truth' about our natural and social worlds. In a sense, multidisciplinary and interdisciplinary research provides a space for all our worldviews to come together, coalesce and extend knowledge.

Finding a 'Community of Practice'

In this concluding chapter, we return to the ROPE group that was mentioned in Chapter 1. As we the editors reflect upon the period when ROPE was active, we conclude that it was a time of watershed moments for the PhD candidates that had opportunities to regularly come together and discuss or debate their research paradigms. There is consensus that 'the concept of research paradigms is one that many higher-degree research students, and even early career researchers, find elusive to articulate, and challenging to apply in their research proposals' (Kivunja & Kunyini, 2017: 26). As mentioned in Chapter 1, in the ROPE meetings, we saw that PhD candidates were confused about several issues relating to paradigms. They raised questions about: what was the concept of paradigms; how to adequately address the concept in their research proposals; how to locate their research in a suitable paradigm; and the level of rationale needed to explain the chosen paradigm.

Other authors also note that such confusion among researchers is not unusual. For example, in the literature there has been emphasis on 'the lingering hegemony of the positivist paradigm, and the dominant perception in the field that high-quality research is associated with quantification' (Klenke, 2016: 336). This thinking was at times evident during the ROPE sessions, and we realised 'the nature of paradigm wars' was alive and well where scholars became steadfast of the hegemony of their respective paradigm and the purists among the participants would resist any adulteration of 'their' paradigms. Such stances highlighted the controversy that is evident in the broader research community (Guba, 1990; Kivunja & Kunyini, 2017) and that even among the ROPE participants it made for lively and polarised conversations that became a feature of the ROPE community of practice.

With such contentious debates in the literature, it is no wonder that participants in ROPE were confused. It became quickly evident, however, that the participants benefitted from conversations on paradigms and the interaction with others who shared challenges in finding their niche in the methodology landscape. Shannon-Baker (2016: 320) notes that 'it is important to engage in discussions about what characterises or can be considered a paradigm'. As an example, she adds 'that "quantitative" and "qualitative" should not be used as synonyms with paradigms' (Shannon-Baker, 2016). It was anomalies such as those pointed out by Shannon-Baker (2016), that were able to be discussed in the 'community of practice' (CoP) to which the ROPE participants belonged.

Such CoPs are defined as a 'learning partnership among people who find it useful to learn from and with each other about a particular domain. They use each other's experience of practice as a learning resource' (Wenger et al., 2011: 9). The original idea of CoP was conceptualised and proposed by Etienne Wenger and Jean Lave in 1991. They maintained that learning is 'a dimension of social practice... [where] learning as increasing participation in communities of practice concerns the whole person... and emphasizes the inherently socially negotiated character of meaning... cognition, learning and knowing' (Lave & Wenger, 1991: 47–50). Later, in 1998, Wenger advanced this concept of CoP in his seminal work, *Communities of Practice: Learning, Meaning, and Identity*. In the book, Wenger focused on workplace learning and argued that people's learnings and identity are shaped by social processes and that a CoP promotes knowing and learning. His ensuing work recognised that CoP is applicable beyond the workplace, e.g. in education. Ultimately, Wenger's idea of CoP has been acknowledged in the conceptualisation of 'social learning'. In his later work, and as seen in the definition above, there is recognition of underlying constructs that characterise the principles of CoP (Wenger et al., 2011). These include: the domain, the community and the practice.

So, it was with ROPE, that participants came together in what Wenger calls 'the domain', i.e. 'the area of knowledge that brings the community together' (Wenger, 2004, para 13), with the community being 'the group of people for whom the domain [i.e. the area of knowledge] is relevant' (Wenger, 2004, para 14). We were all wanting to learn more about paradigms (P), their ontologies (O) and epistemologies (E), and their place in research (R). Hence, the acronym for the group – bringing all of these aspects together under the auspice of research. With hindsight, we could have adopted the acronym of ROPEM, especially as the participants mastered (M) the philosophies of research. Through 'the practice' we were able to build our knowledge and expertise. Wenger (2004) says practice is 'the body of knowledge, methods, tools, stories, cases, documents, which members share and develop together' (para 15). Our regular ROPE meetings allowed for providing a common focus (domain), building of relationships to foster learning (community) and securing the learning by bringing what we were learning to the writing of our respective PhDs and the practice of research. These collective and individual activities were possible because of the cultivating of a CoP.

Being a Reflective Practitioner

Hall (2004: 150) made the comment that, 'as we peel off the onion-like layers of our understanding concerning the nature of our research', we need to be serious about reflexivity and ask ourselves, 'how do we situate our research with ourselves and others before we go and observe and involve ourselves with those others'. We need to understand our positionality, personal values and basic assumptions during the scoping and design of our research projects as well as when we are in the field. For the authors in this book and the participants in ROPE, the exploration of paradigms, ontologies and epistemologies has demanded that they, as scholars, examine their role in the research process and to be aware of the biases that may permeate into their research designs and processes. Invariably, reflexivity is generally associated with researchers who are utilising qualitative methods, where it is used to recognise the potential effects of the research on the research participants and vice versa. However, at the bare minimum, even the positivists among us, those who maintain an objectivist ontological position, could limit their reflexivity to 'a localised critique and evaluation of the technical aspects of the particular methodology deployed rather than the underlying metatheoretical assumptions that justify that methodology in the first place' (Johnson & Duberley, 2003: 1284).

Reflexivity is defined as, 'an awareness of the relationship between the investigator and the research environment' (Lamb & Huttlinger, 1989: 766). It mandates that the researcher reflects on how s/he will be influencing the research environment and how s/he will be (or are or has been) influenced by the research environment. It is inescapable that the nature of research warrants that the researcher will become part of the research, even if the intent is to maintain objectivity. Some authors argue that 'there is something to be gained in acknowledging and dealing with the investigator's intimate relationship with the research study' (Lamb & Huttlinger, 1989: 767). Walker *et al.* (2013: 42) reported that keeping a reflective research diary 'provided the researcher with instantaneous opportunities to reflect on the issues relating to process and content associated with collecting data'. So, the value of the reflections extends beyond authenticating the validity and reliability of the data; the reflections present opportunities to 'improve and inform future research practice' (Walker *et al.*, 2013: 42). Chapter 9 of this book includes an excerpt of a reflective research diary kept during data collection. It can be thought of as best practice for researchers of all paradigms to maintain a reflective research diary that not only allows the researcher to record information about the research process but also to provide a transparent source of the journey. Walker *et al.* (2013) emphasise the relevance of having a reflective research diary for both quantitative and qualitative approaches. They refer to the 1992 writings of Joseph Maxwell and his contention that, 'In these postmodern times, it is rarely asserted that even hard science research can produce absolute certainty' (Walker *et al.*, 2013: 43). Harré (2004) asserts that imparting of the researcher's positionality through reflexive practice affords meaning to the research and specificity to the results and that through this process, it enhances the scientific value of the research. Hence, maintaining a reflective research diary provides a rich source of day-to-day activities, challenges and discoveries and lends insight into any queries about the research that may arise in due course. Critical to this process is the reflexivity.

Haynes (2017: 290) states that, 'the process of enacting reflexivity in research is sometimes daunting'. She refers to the work of Hibbert *et al.* (2010 cited in Haynes, 2017) and outlines that the reflexivity process includes four steps – repetition, extension, disruption and participation – that integrate reflection and recursion. For example, with the first step of 'repetition', individuals deliberately focus on continuous reflection, such as what are their underlying assumptions and how do these influence their research design. In the latter steps self-awareness and self-critique become more pronounced and during these steps, 'basic assumptions and values are challenged, and ultimately potentially transformed. It is in such moments that . . . [the researcher may be] questioning both self and knowledge' (Haynes, 2017: 291). Hence, through this process, researchers are forced to examine their position as it relates to theory, methodology, participants and themselves.

For the scholars exploring paradigms, reflexive practice is important for contextualising the research and ensuring that the researcher understands their position in relation to the theoretical stances taken at the ontological, epistemological and methodological levels. Haynes (2017: 294) sums up to say that:

> . . . reflexivity questions the processes and practice of research, in terms of how our methodological conduct and theoretical pre-understandings as researchers transform and influence new understandings . . . Researchers need to be reflexively aware of how their pre-understandings influence the design and conduct of their research and how they are influenced by the process of research itself. Hence, everything from the choice of topic, research question, research design, methodology and theoretical interpretation should be subject to reflexive questioning on the influence of the researcher, as well as the influence of the research on the researcher.

Finlay (1998: 453) proffers that reflexivity can reveal 'new understandings'. Palaganas *et al.* (2017: 426) add that conducting research and in particular, the undertaking of fieldwork, changes the researcher. Through the ritual and discipline of being reflexive, the associated notes and insights can contribute to the research findings and 'the constructions of meanings'.

In 1999, Karl Weick talked of 'the reflexive turn' and how as researchers, we must connect with ourselves by reflecting on our own thinking throughout the research process. Burrell and Morgan (1979) were among the first authors who gave assent to the 'newspeak', as Willmott (1993) described it. They upheld that researchers' paradigms heightened the proclivity of researchers to respective choices that were made throughout the research process. Others added to the discussions by saying that 'while we cannot eradicate our subjective metatheoretical commitments, we must open them to our inspection through our capacity for reflexivity' (Johnson

& Duberley, 2003: 1280). Such was the thinking of scholars such as Bourdieu and Wacquant (1992: 68) who saw reflexivity as a 'fundamental dimension of epistemology'. Through ROPE, we sought to engender reflexivity and reflective practice. This book is testimony to the animated conversations that challenged us to think about our own thinking and the default stances we took into our research. Our work is richer for the experiences of being able to here articulate those reflections and share our new gained knowledge with you, the reader.

Conclusion

The chapters in this book give evidence to how paradigms can be used, some of the challenges in understanding the jargon and technicalities of paradigms, and the importance of explicitly engaging with paradigms. Each author has shown how a paradigm can be operationalised and has discussed some of the dilemmas in doing so. We hope that this book will assist emerging scholars in reporting on their paradigms and their experiences in using their paradigms. Such discussions will enhance our understanding of the conceptualisation, characteristics and mechanics of working with different worldviews in our research. Knowledge of the philosophical underpinnings of our research enables us to be better positioned to understand how we comprehend, design, analyse and interpret the research which will inform the contributions that we make to new knowledge.

Natural and social dilemmas as well as human curiosity direct researchers to make enquires of the world we live in. The rich and diverse contexts of our everyday environments present settings within which to examine these dilemmas. The researchers' own sensitivities and underlying basic assumptions inform the lens through which the research is being conducted. Awareness of one's own extensions and limitations in thinking and being is essential to the research process. One's maturation as a researcher demands engagement with discussions around paradigms, ontologies, epistemologies and methodologies. For the future, we hope that scholars will continue to engage in lively and challenging discussions on the seemingly esoteric aspects of paradigms, within which are embedded the paradoxes and dilemmas of *Life, the Universe and Everything*, to draw from the title of Douglas Adams' infamous 1982 book, which is the third volume in a series of five books that are part of his *Hitchhiker's Guide to the Galaxy*.

References

Adams, D. (1982) *Life, the Universe and Everything*. London: Pan Books.
Allsop, J. (2012) Competing paradigms and health research: Design and process. In M. Saks and J. Allsop (eds) *Researching Health: Qualitative, Quantitative and Mixed Methods* (pp. 18–41). London: Sage Publications Ltd.

Bourdieu, P. and Wacquant, L.J.D. (1992) *An Invitation to Reflexive Sociology*. Chicago: University of Chicago Press.

Bryman, A. (2006) Paradigm peace and the implications for quality. *International Journal of Social Research Methodologies* 9 (2), 111–126.

Burrell, G. and Morgan, G. (1979) *Sociological Paradigms and Organizational Analysis*. Oxford: Heinemann.

Finlay, L. (1998) Reflexivity: An essential component for all research? *The British Journal of Occupational Therapy* 61 (10), 453–456.

Guba, E. (1990) (ed.) *The Paradigm Dialog*. London: Sage Publications Ltd.

Hammersley, M. (1992) The Paradigm Wars: Reports from the front. *British Journal of Sociology of Education* 131 (1), 131–143.

Hall, M. (2004) Reflexivity and tourism research: Situating myself and/with others. In: J. Phillimore and L. Goodson (eds) *Qualitative Research in Tourism: Ontologies, Epistemologies, and Methodologies* (pp. 137–155). Hove: Psychology Press.

Harré, R. (2004) Staking our claim for qualitative psychology as science. *Research in Psychology* 1 (1), 3–14.

Haynes, K. (2017) Reflexivity in accounting research. In Z. Hoque et al. (eds) *The Routledge Companion to Qualitative Accounting Research Methods* (pp. 284–298). Abingdon: Routledge.

Johnson, P. and Duberley, J. (2003). Reflexivity in management research. *Journal of Management Studies* 40 (5): 1279–1303.

Klenke, K. (2016) Epilogue. In K. Klenke (ed.) *Qualitative Research in the Study of Leadership* (2nd edn, pp. 331–354). Bingley: Emerald Group Publishing.

Kivunja, C. and Kuyini, A. (2017 Understanding and applying research paradigms in educational contexts. *International Journal of Higher Education* 6 (5), 26–41.

Lamb G. and Huttlinger K. (1989) Reflexivity in nursing research. *Western Journal of Nursing Research* 11 (6), 765–772.

Lave, J. and Wenger, E. (1991) *Situated Learning: Legitimate Peripheral Participation*. Cambridge: Cambridge University Press.

Ling, L. and Ling, P. (2016) Conclusion: Paradigm and paraddidle. In P. Ling and L. Ling (eds) *Methods and Paradigms in Education Research* (pp. 345–351). New York: IGI Global.

Mackenzie, N. and Knipe, S. (2006) Research dilemmas: Paradigms, methods and methodology. *Issues in Educational Research* 16 (2), 193–205.

Merriam, S.B. and Grenier, R.S. (2019) Assessing and evaluating qualitative research. In S.B. Merriam and R.S. Grenier (eds) *Qualitative Research in Practice: Examples for Discussion and Analysis* (pp. 19–32). Chichester: John Wiley & Sons.

Mertens, D.M. (2005) *Research Methods in Education and Psychology: Integrating Diversity with Quantitative and Qualitative Approaches* (2nd edn). London: Sage.

Palaganas, E., Sanchez, M, Molintas, V., Caricativo and Ruel, D. (2017) Reflexivity in qualitative research: A journey of learning. *Qualitative Report* 22 (2), 426–438.

Smith, J.K. (1989) *The Nature of Social and Educational Inquiry: Empiricism Versus Interpretation*. Boston: Ablex Publications.

Shannon-Baker, P. (2016) Making paradigms meaningful in mixed methods research. *Journal of Mixed Methods Research* 10 (4), 319–334.

Tribe, J. (2004) Knowing about tourism: Epistemological issues. In J. Phillimore and L. Goodson (eds) *Qualitative Research in Tourism: Ontologies, Epistemologies, and Methodologies* (pp. 46–62). Hove: Psychology Press.

Walker, S., Read, S. and Priest, H. (2013) Use of reflexivity in a mixed-methods study. *Nurse Researcher* 20 (3), 38–43.

Weick, K.E. (1999) Theory construction as disciplined reflexivity: Trade-offs in the 90s. *Academy of Management Review* 24 (4), 797–810.

Wenger, E. (1998) Communities of practice: Learning as a social system. *Systems Thinker* 9 (5), 2–3.

Wenger, E. (2004) Knowledge management as a doughnut. *Ivey Business Journal*. Retrieved from: http://iveybusinessjournal.com/publication/knowledge-management-as-a-doughnut/

Wenger, E., Trayner, B. and de Laat, M. (2011) *Promoting and Assessing Value Creation in Communities and Networks: A Conceptual Framework*. Amsterdam: Open University of the Netherlands.

Willmott, H.C. (1993) Strength is ignorance; slavery is freedom: Managing culture in modern organizations. *Journal of Management Studies* 30 (4), 515–52.

Index

Note: References in *italics* are to figures, those in **bold** to tables.

abduction 162, 165
Abercrombie, N. *et al.* 182
Abreu-Novais, M. *et al.* 145
academic tribes 204
Acker, J. 135, 136
action research 194
Adams, D. 210
Allen-Brown, V. 114
Allsop, J. 205
Alvesson, M. 133
Amazon's Mechanical Turk (AMT) 21–22
Anderson, A. 2
Andrades-Caldito, L. *et al.* 145, **146**, 147, *147*
Arnould, E.J. 116, 120
Asian spas *see* intermediate paradigmatic ground
Atkinson, P. 103, 106
Aubert-Gamet, V. 121
autoethnography, defined 97
 see also ethnographic exploration of hotel work and hospitality in far north Queensland
axiology 6, 90, 177, **179**, **181**, 183, 191, 192, 193–194, 210

Babbie, E. 92
Baldwin, S. 40
Banai, R. 163
Barley, S. 103
Barnes, T.J. 161
behavioural science 26
Berger, P. 65
Berkeljon, A. 40
Betzing, J.H. *et al.* 28
Bishop, J. 132, 133
Blau, P. 107
Bloom's taxonomy 191–192
Boas, F. 102
Boghossian, P. 195
Bosnia-Herzegovina: cross-cultural identity 72–78, **76**, **78**
Botterill, D. 177
Bourdieu, P. 210
Bowling, A. 102
Boyle, A.R. 93
bracketing 69–70, 71, 74
Bryman, A. 205
Burrell, G. 209

Calder, B.J. *et al.* 19
Cambridge Dictionary 159
Campbell, D.T. 19
Carmona, M. *et al.* 162, 163, 164, 166, 167–168
Cassidy, L. 33
Castleden, H. *et al.* 52
causal relationships 12–17
 in experimental designs 18–19
CBTEs *see* community-based tourism enterprises
chapter structure 3–4
Charles, G. 102–103
Chilisa, B. 6
Clawson, M. 174
Cobb-Stevens, R. 69
Coghlan, A. 145, **146**, 146–147, *147*
Cohen, E. 196
Cohen, S. 196
Coles, T. *et al.* 161
Collier, K.G. 90
communities of practice 206–207
community-based tourism enterprises (CBTEs) 50–51, 55
 see also knowledge co-production in tourism
confirmation-of-candidature 2
constructivism 51, **178–179**, 191, 193, 195, 199

constructivist paradigm and
    phenomenological research
    design 64–65
  application: cross-cultural identity
    72–78, **76**, **78**
  bias 67
  epoché 69–70, 71, 74–75
  phenomenology 64, 65, 68–71
  reality 66
  reflection on use of paradigm 78–80
  review of the paradigm 65–71
  conclusion 80–81
  see also knowledge co-production in
    tourism
consumer behaviour *see* logical
    positivism in consumer
    behaviour research
content analysis 92
convenience sampling 45
Corbin, J. 67
Corley, K.G. 200
Cova, B. 120, 121
Cova, V. 120
Creswell, J..D. 100
Creswell, J.W. 7, 45, 67, 100
critical realism 163, 177, **178–179**, 183
critical theory 114–115, **178–179**, 191,
    193–195, 199–200
  see also neo-tribalism through an
    ethnographic lens
Cronbach's Alpha 23
cross-cultural identity in Bosnia-
    Herzegovina 72–78, **76**, **78**
cross-disciplinary, defined 160
Crotty, M. 78
Crouch, G. I. 145, **146**, 147, *147*, 148
Csikszentmihalyi, M. 175

Dalton, M. 107
Darbellay, F. 88
Davids, T. 107
Davis, T.R. 101
deduction 162
Denis, A. 133, 134
design science research methodology
    (DSRM) 27–29
  activities 28, *28*
  website analysis method (WAM)
    30–32, *31*, *32*, *34*
design science research (DSR) paradigm
    25–29

application 28–33
defined 26
guidelines 26–27
final thoughts 33–36
destination competitiveness *see*
    pragmatic paradigm in
    destination competitiveness
    studies
Devitt, M. 66
Dewey, J. 143
Dharamsi, S. 102–103
D'Hauteserre, A.-M. 145
Disch, L. 131, 132
disciplinary research 160–161, *161*
Dorst, K. 162
Dos Santos, M.C. 117
Doucet, S.A. *et al.* 115
Dovey, K. 163
Druckman, J.N. *et al.* 19, 22
DSRM *see* design science research
    methodology
Duberley, J. 208, 209–210
Duham-Quine thesis 16
Durkheim, É. 116
Dwyer, L. 145, 146, **146**, *147*

E Silva, S.C. 117
Easterby-Smith, M. 14, 15
Echtner, C.M. 164
El-Masri, M. 29
Ellis, C. *et al.* 97
Ellis, G.D. *et al.* 175
Ely, R.J. 137–138
emic perspective 72, 80, 92, 103, 175,
    183–184
epistemology 6, 84, 88–89, 177, **178**,
    181, 182, 191, 192
epoché 69–70, 71, 74–75
ethnographic exploration of hotel work
    and hospitality in far north
    Queensland 97–98
  ethnographic approach 101, 102–105
  ethnographic techniques 105–106
  interpretivism in PhD research
    100–102
  journey to find my paradigm
    98–100
  organisational culture (OC) 97,
    100–101
  qualitative approach 99, 100, 101
  reflecting on my journey 106–108

triangulation 97, 98, 100
  conclusion 108–109
ethnography, defined 102
  see also neo-tribalism
etic perspective 175, 183–184
experimental designs
  between- and within-subjects designs 21
  causality 18–19
  data collection strategy 21–22
  factorial designs 20, **20**
  one-way designs 20
  reflections and elaboration 22–23
  reliability 23
  validity 19–20, 22, 23, 40
external validity 19, 22, 23

factorial designs 20, **20**
falsificationism 14–15, 16, 19
Faulconbridge, J. 196
Feilzer, M.Y. 109
feminisms 130, 194
  application of the paradigm 136–138
  feminist methodologies 135–136
  femocrats 132
  gender mainstreaming and diversity management 134
  identity politics 132–133, 134
  inequality regimes 135
  intersectionality 134–135
  liberal feminism 132
  neoliberal feminism 133–134
  recent developments and ways of looking 133–135
  reflection on use of the paradigm 138–139
  review of the paradigm 131–136
  waves of feminism 131–133
  conclusion 139–140
Fereday, J. 151
Fetterman, D. 107, 183
field experiments 42
fields of study 204
Filep, S. 175
Finlay, L. 209
Fisher, R. 1, 5
Fraser, N. 133
functional magnetic resonance imaging (fMRI) 41
future: moving forward with reflective practice on paradigms 203–204
  being a reflective practitioner 207–210
  finding a 'Community of Practice' 206–207
  multidisciplinary and interdisciplinary research 204–206
  conclusion 210

Gabott, M. 148
Gale, T. 177
Gardial, S.F. 148
gay tourism *see* neo-tribalism through an ethnographic lens
Gearing, R. 69
Gelo, O. 18
gender mainstreaming and diversity management 134
Gioia, D.A. 200
Giorgi, A. 68
Giorgi, B. 68
Goffman, E. 107
Gold, R. 106
Goldstein, H. 41
Goodson, L. 180
Gould, S.J. 195
Graul, A. 17–18
Greenacre, L. *et al.* 124
Grenier, R.S. 203
Guba, E.G. 2, 5, 51, 114, 164, 177, 182, 183, 205

Hair, J.F. *et al.* 45
Hall, C.M. 167–168
Hall, M. 207
Hamilton, J. 32
Hammersley, M. 103, 106, 205
Hardy, A. *et al.* 115–116, 118, 119, 122
Harré, R. 199, 208–209
Harrill, R. 180
Hassan, S.S. 145
Hawkesworth, M. 131, 132
Haynes, K. 209
Heidegger, M. 68
Hemmings, C. 133
hermeneutics 66, 68
Hevner, A.R. *et al.* 26, 28, 30, 35
Hogg, G. 148
Hollinshead, K. 176, 177
Hong, W.C. 145, **146**, 147, *147*
Hookway, C. 159
Horkheimer, M. 114

hotel work and hospitality *see* ethnographic exploration of hotel work and hospitality in far north Queensland
Howe, K.R. 154
Hughson, J. 120–121, 124
Hui, A. 196
Hume, D. 13–14
humour *see* quasi-experiments to study humour in tourism settings
Husserl, E. 65, 68, 69, 71, 80
Huttlinger, K. 208
hypotheses 15, 16

identity: cross-cultural identity in Bosnia-Herzegovina 72–78, **76, 78**
identity politics 132–133, 134
ideology 84, 90
idiographic approach 72–73, 101
induction 14, 162
inequality regimes 135
information systems (IS) 25, 26, 27
information technology (IT) 26
interdisciplinarity
　defined 160
　fields 160, 161
　research 204
　see also pragmatism in context of urban design and tourism
intermediate paradigmatic ground: Asian spas 171–172, **181**
　axiology 177, **179, 181**, 183
　concepts 174–175
　emic and etic perspectives 175
　epistemology 177, **178, 181**, 182
　major research paradigms 177, **178–179**
　methodology 177, **179, 181**, 182–183, 183
　mixed paradigms in spa tourist motivations and experiences 180–184
　ontology 177, **178**, 180, **181**
　paradigms in tourism 176–184
　PhD: overview and context 173–175
　positive psychology 175
　post-positivism and critical realism 177, 180, **181**
　tourist behaviour and motivation 174–175
　Travel Career Pattern (TCP) theory 175
　wellness 174
　post-PhD musings on paradigms 184–185
internal validity 19, 22, 23
interpretive social science paradigm 84–85, 85
　epistemology 84, 88–89
　ideology 84, 90
　my story 85–87
　ontology 84, 87–88
　reflections on application 92–93
　review of the paradigm 91
　tourism education 84
　conclusion 93
　see also ethnographic exploration of hotel work and hospitality in far north Queensland
interpretivism 98, 99, 100–102, **178–179**, 193
interpretivist qualitative thinking 8
intersectionality 134–135
intradisciplinary, defined 160
intuition 69
Ivanov, S. 90

Jafari, J. 91
Jamal, T. 5, 177
Jennings, G. 87, 176, 177, 182
Johnson, P. 208, 209–210
Johnson, R.B. 9

Kant, I. 67
Kao, H.-Y. *et al.* 28
Kaushik, V. 143
Kawulich, B. 6
Kelly, C. 174
Khoo-Lattimore, C. *et al.* 142–143
Killion, L. 1, 5
Kim, C. 145, 146, **146**, *147*
Kim, S.S. *et al.* 152
Kindon, S. *et al.* 57, 61
Kivunja, C. 206
Klenke, K. 206
Knetsch, J.L. 174
Knipe, S. 91, 204
knowledge co-production in tourism 50–51
　application: research context (Vietnam) 55–59

community-based tourism enterprises (CBTEs) 50–51, 55–59, **56**
constructivism as research paradigm 51
PAR (participatory action research) 53–55, **54**
PAR as methodological framework 55
reflexivity 59–61
research design 57, 57–59
in tourism research 52–54
conclusion 61
Kostera, M. 107
Krejcie, R.V. 152
Krishnan, A. 160, 161–162
Kuechler, W. 27
Kuhn, T.S. 4, 15–16, 176, 177, 185, 190, 191
Kunyini, A. 206

ladenness of observation theory 16
Lakatos, Imre 16–17
Lamb, G. 208
Lave, J. 207
Le, D. *et al.* 144
LeCompte, M.D. 122–123
Lee, A.S. 98–99
Lee, R.M. 42
Lee, U.-I. 175
Lee-Ross, D. 100
Leech, N.L. 8, 9, 165
Legg, C. 159
Leiper, N. 87, 160
Lewis, D. 14
Liburd, J.J. 88
Lichtman, M. 74
Lincoln, Y.S. 2, 51, 64, 164, 177, 183
Lindsay, J. 195
Ling, L. 203
Ling, L.M. 84
Ling, P. 203
logical positivism 14
logical positivism in consumer behaviour research 12
application in social science research 17–22
between- and within-subjects designs 21
causal relationship 12–17
causality in experimental designs 18–19
data collection strategy 21–22
historical and philosophical tenets 13–17
Hume's theory of causation 13–14
internal/external validity in experimental designs 19–20, 22
Kuhn's critique 15–16
Lakatos' response 16–17
one-way and factorial designs 20
Popperian falsificationism 14–15, 22
reflections and elaboration of experimental designs 22–23
validity in experimental designs 19–20
Loukaitou-Sideris, A. 163–164
Luckmann, T. 65
Luthans, F. 101

MacFarlane, J. 194
McGregor, S.L.T. 5, 39
Mackenzie, N. 84, 91, 204
McKercher, B. 90
McKnight, L. 130
McRobbie, A. 130
Maffesoli, M. 116, 117
Maitland, R. 163
Malhotra, N.K. *et al.* 18, 22
Malinowski, B. 102, 106
Martin, J. 103
Mason, J. 123
Maxwell, J. 208
Mayo, E. 107
Mazanec, J. 195
mental models 27
Merriam, S.B. 203
Mertens, D.M. 5, 204
methodology 6, 8, 9, 177, **179**, *183*, 191, 192
methods, defined 8
#metoo 130, 132
Meyerson, D.E. 137–138
Mill, J.S. 14
mixed-methods research 143–144, 205–206
Mobilities 196
Morgan, D. 142, 159
Morgan, D.L. 159, 162, **162**
Morgan, D.W. 152
Morgan, G. 209
Morris, I. 197
Moustakas, C. 69–71, **76**, 81
Muir-Cochrane, E. 151

multi-paradigmatic research 5, 8–9
multidisciplinarity, defined 160
multidisciplinary, defined 160
multidisciplinary research 204–206
Munar, A.M. 5, 177
Murnane, J.A. 5, 38
Myers, M.D. 124

neo-tribalism 115–117
neo-tribalism through an ethnographic lens 112–114, 208
   application of neo-tribalism and ethnography 121–124
   characteristics of neo-tribalism 119–121, *120*
   critical theory paradigm 114–115
   neo-tribalism and tourism 118–119, **119**
   reflections 124–125
   conclusion 125–126
Neyland, D. 107
Nichols, R.G. 114
#notallmen 130

ontology 6, 84, 87–88, 159, 177, **178**, 180, **181**, 191, 192
Onwuegbuzie, A.J. 8, 9, 165
O'Reilly, K. 106
organisational culture (OC) *see* ethnographic exploration of hotel work and hospitality in far north Queensland
Orr, J. 107
otherness 86
Ott, S.J. 102

Pabel, A. 2, 46, 145, **146**, 146–147, *147*
Painter, J. 161
Palaganas, E. *et al.* 209
Panchal, J.H. 183
paradigms 1, 2–3
   application 4
   author's use of 4
   choosing the 'correct' paradigm 6–7
   defined 4–5, 15, 16
   discipline-specific influence 6
   personal considerations 7
   principles 5–6
   research design 7–9
   review 3
   supervisor influence 7
   in tourism 176–184
   conclusion 9–10
paradigms: a supervisor's perspective and advice 189–190
   current situation 199–200
   fundamental positions 192–196
   new paradigm candidates 196–197
   personal encounters 197–199
   a special language 190–192
   conclusion 200
participatory action research *see* knowledge co-production in tourism
Patton, M.Q. 100–101
Pearce, D. 171
Pearce, P.L. 5, 46, 172, 173–174, 175, 176, 205
Peck, J. 167–168
Peffers, K. *et al.* 27–28, 30, 35
Peterson, J. 195
phenomenology 64, 65, 68–71
   see also constructivist paradigm and phenomenological research design
Phillimore, J. 180
philosophical framing 2, 198
Pierce, C.S. 159
Pisani, E. 193
Pluckrose, H. 195
Ponterotto, J.G. 39
Popper, Karl 14–15, 22
positivism 175, 176, **178–179**, 191, 192, 199
positivist quantitative thinking 8, 98–99
post-disciplinary approach 161–162
post-positivism 15, 38–40, 176, 177, **178–179**, 180, **181**, 183, 191, 193
   see also quasi-experiments to study humour in tourism settings
Potts, T.D. 180
Powell, T.C. 143
Pragmatic Maxim 159
pragmatic paradigm in destination competitiveness studies 142
   interviewing SCUBA divers 148, 150–151, **151**
   methodological framework 152, *153*
   mixed-methods approaches 143–144
   models 145–146, **146**, *147*

online survey 151–152
research design 148–154, *149*
review of pragmatic paradigm 142–144
reflection on use of the paradigm 154
pragmatism 109, 159, **178–179**, *183*
pragmatism in context of urban design and tourism 158–159
multidisciplinary study 160–162
my research: critical pragmatic case study 163–166, **164**
post-disciplinary study 161, **162**, 162–163
pragmatic approach 159–160
reflection on pragmatic approach 166–168
conclusion 168
Prasad, T.B. 164
Prideaux, B. 90
problem of induction 14–15
Pryce, J. 2
public health research 204–205
purists *183*

qualitative thinking 8, *183*
quantitative thinking 8, 18, 98–99, *183*
quasi-experiments to study humour in tourism settings 37
application of paradigm in PhD research 40–46
manipulation of treatment scenarios 44, **45**
method 39, 41–42
naturalistic quasi-experimenting 42–43
on-site procedures 44–45
questionnaire data analysis 45–46
reflections on experience 46–47
review of post-positivist paradigm 38–40
tourism settings and questionnaire design 43–44
validity 40
conclusion 47–48
queer theory 194
Queiroz Neto, A. *et al.* 145, 151, 154

reality 66
reason 15
Reeves, S. *et al.* 101, 102, 106

reflective practitioners 207–210
reflexivity 4, 59–61, 208–210
Reiser, A. 47
relativism 194
reliability in experimental designs 23
Renault, C. 41
research design 7, 7–8
multi-paradigmatic research 8–9
research paradigms 51, **178–179**, 199
Richard, L. 92
Ritchie, J.R.B. 145, 146, **146**, *147*, 148
Rittel, H.W. 168
Robards, B. 118, 123
Robson, C. 4
ROPE group 2–3, 206, 207, 208, 210
Rossman, G.B. 182
Rottenberg, C. 133–134
routine design 26
Ryan, C. 199

Sandberg, S. 133
Saracci, R. 174
Sarantakos, S. 185
Schein, E. 101, 109
Schein, R. 131
Schensul, J.J. 122–123
Schutz, A. 98
scientific research programs 16
SCUBA diving tourism *see* pragmatic paradigm in destination competitiveness studies
Seligman, M.E.P. 175
sensationalist perspective 183, *183*
Shannon-Baker, P. 205, 206
Shearer, M.C. 89
Shenton, A.K. 60
Sholle, D. 161
Siehl, C. 103
Silverman, D. 107
Simmons, D.G. 47
Simon, H.A. 26
Skóldberg, K. 133
Smith, A. 163
Smith, D.W. 71
Smith, J.K. 205
Smith, M. 174
social justice 114
social learning 207
social science research: logical positivism 17–22
Sokal, A.D. 195

Sousa, V.D. *et al.* 5
spas *see* intermediate paradigmatic ground: Asian spas
Spasojevic, B. *et al.* 144
Stanley, J.C. 24
Stember, M. 160
Stevick–Colaizzi–Keen method 75, **76**
Stock, M. 88
Stockmann, N. 177
Stratton, G. 194
Strauss, A. 67
subjectivism 182
Sung, H.H. 152

Tarhini, A. 28
Tashakkori, A. 176
Teddlie, C. 176
Tedlock, B. 121–122, 123
Teubner, T. 17–18
Theodore, N. 167–168
Thompson, C.J. 116, 120
Ticehurst, B. 102
tourism *see* intermediate paradigmatic ground; knowledge co-production in tourism; neo-tribalism through an ethnographic lens; pragmatic paradigm in destination competitiveness studies; pragmatism in context of urban design and tourism; quasi-experiments to study humour in tourism settings
tourism education *see* interpretive social science paradigm
transcendental I 71
transdisciplinary, defined 160
Travel Career Pattern (TCP) theory 175
Tribe, J. *et al.* 87, 88, 90, 91, 177
Tunnell, G.B. 42–43
Turner, V.W. 119

urban design *see* pragmatism in context of urban design and tourism
Urry, J. 196

Vaishnavi, V. 27
validity in experimental designs 19–20, 22, 23, 39
Van Maanen, J. 107
Veal, G.W. 102
Vietnam *see* knowledge co-production in tourism
Vorobjovas-Pinta, O. 121, 122, 123

Wacquant, L.J.D. 210
Walby, S. 134
Walker, S. *et al.* 208
Walle, A.H. 177
Walsh, C.A. 143
Webber, M.M. 168
Weber, M. 91
website benchmarking *see* design science research (DSR) paradigm
Webster, C. 90
Weick, K. 209
Weick, K.E. 42
wellness 174
Wenger, E. *et al.* 207
White, P.A. 13
Whyte, W.F. 103, 107
Wilde, S.J. 145, 146, **146**, 147
Willis, J.W. 39
Willmott, H.C. 209
Wilson, B.L. 182
Wollstonecraft, M. 131
Woodruff, R.B. 146, 147, 148

Ybema, S. *et al.* 107
Yin, R.K. 165

Zammito, J.H. 39